Noncapitalist Development

This study traces the socio-political development that culminated in the triumph of the Marxist-Leninist program to "resist imperialism" and to place the profitable sugar industry firmly in the hands of "the people." The resulting national unity and euphoria were short-lived, according to the author, and today the working class remains divided along racial lines, Marxism-Leninism is reduced to empty rhetoric, and the government maintains its rule through repression, corruption and racial ploys. This book contains many lessons for understanding the problem of development in the noncapitalist Third World.

Noncapitalist Development

The Struggle to Nationalize the Guyanese Sugar Industry

Paulette Pierce

ROWMAN & ALLANHELD
PUBLISHERS

ROWMAN & ALLANHELD

Published in the United States of America in 1984
by Rowman & Allanheld, Publishers
(A division of Littlefield, Adams & Company)
81 Adams Drive, Totowa, New Jersey 07512

Library of Congress Cataloging in Publication Data
Pierce. Paulette.
  Noncapitalist development.
  Bibliography: p.
  Includes index.
  1. Sugar trade—Government ownership—Guyana.
I. Title.
HD9114.G82P54 1984   338.1'8   84-17882
ISBN 0-86598-118-3

84 85 86 / 10 9 8 7 6 5 4 3 2 1
Printed in the United States of America

TABLE OF CONTENTS

Acknowledgements

Since this work emerged from my doctoral dissertation I'd like to once again thank the members of my committee: Dr. Bogdan Denitch, Dr. Frank Bonilla, Dr. Raymond Franklin and Dr. Stanley Aronowitz. Their encouragement and honest criticism during and since that time has been an invaluable resource in my struggle to understand the problems of development in the Third World. Private corporations are not known for their openness. Therefore, special thanks are due to the members of the London Office of Booker McConnell, Ltd. who so freely shared their insights and allowed me access to many important company records. Lastly, my deepest gratitude is due to the many individuals in Guyana who without thought of personal gain or security cooperated with this research endeavor and in large measure made it possible.

Introduction

The center of the development debate after the Second
World War was whether colonialism was the basis for third
world poverty and underdevelopment. For those who an-
swered in the affirmative, movements for "national liberation" whose primary objective was to achieve political
independence were, perhaps unwittingly, equated with
development. A second issue--the form of the state--grad-
ually displaced the concern with independence after the
1950s when the leading metropolitan countries, particularly Great Britain and France, were forced to divest
themselves of direct political control over Asian, African and Latin American countries. Modernization theorists such as Seymour Martin Lipset, S.N. Eisenstadt and W.
Arthur Lewis argued that liberal democracy within private
ownership was both the end and the best means for achieving economic and social development and freedom. They
generated a set of criteria, including the doctrine of
comparative advantages that effectively tied third world
development to the metropolitan countries even after
political independence had been secured. The two left-
wing alternatives were leninism and dependency theories.
The leninist critique of imperialism tried to demonstrate
that western democratic forms were impossible in the
context of third world development because the fundamental
precondition--a national bourgeoisie and a free labor
movement--were not yet present. According to this mode of
thought, the tasks of the national liberation movement
were first, to expel the imperialists from the country;
second, to create a state forum that stimulated the
primitive capital accumulation process; and third, to
cement the alliance between the peasantry (by far the
majority in nearly all third world countries), the urban
proletariat, and the revolutionary intellectuals and
middle classes to secure state power in the long-range
benefit of their respective sectors. According to this
theory/program, the newly independent state is obliged to
perform the role of capitalist without extensive private
ownership and control of the decisive means of production.
In order to achieve the basic objective of economic as
well as political independence from the former colonial
powers, democracy would have to be redefined; hence some
version of the "democratic dictatorship" of the nascent
proletariat and the peasantry under the leadership of
a marxist-leninist party which could gradually change its
composition from that of primarily intellectuals to a
broadly based force. A variant on this development
paradigm was the military as the democratic dictatorship.
In this case, the party and the revolutionary military

collaborate to secure popular power through either the
anti-imperialist or civil war. Such instances as Vietnam,
Mozambique and Algeria might be considered examples of
this model which, in any case, is dictated by the refusal
of the colonial powers to grant political independence
without significant armed struggle.

The dependency theory argues that there is no necessary connection between political independence and economic and social autonomy, for the strings that tie the third
world to erstwhile colonial powers are the chronic need
for investment capital. Inevitably, insist the dependency
theorists, the state of a conventional bourgeoisie must
turn to the western powers for capital, since the Warsaw
pact countries are simply incapable of providing such
resources. Even if the state is the main capitalist, the
bonds of dependency are secured by the World Bank, the
International Monetary Fund or private investment capital
from the leading financial centers, which demand control
as a price for loans. Anyone familiar with the recent
debt crisis in such countries as Mexico, Brazil, Argentina
and Jamaica will find this theoretical perspective played
out in the newspapers every day. Apparently, political
independence does not insure economic self-determination
unless the country in question chooses to sever its ties
with western capital and embarks on a long march to
industrialization through the virtual militarization of
the economy. In any case, only the Soviet Union and China
have sustained such an effort, and as their present
governments acknowledge, at substantial cost in human
hardship. This "dissociation" model has never existed in
pure form; even the Soviets and Chinese have periodical-
ly backed away from it, especially on such issues as
technology transfer and, more recently for China, direct
capital investment. Critics of the Soviet path to economic development have shown that the Soviet rhetoric of
independent economic development has never been matched
by practice. The Soviets have always relied on western
technology and, in the 1920s, capital inputs from private
sources. Consequently, as Frederick Fleron has argued,
the Soviet system is, in many decisive respects, similar
to that of the capitalist west, even if their political
systems differ.

These perspectives have often been accompanied by
mutually exclusive claims. In their concern to
establish intellectual hegemony each has emphasized
its respective differences, rather than similarities.
And, in the plethora of case studies that support each
view, the others are shown to be inapplicable to the
specific situation. Unfortunately, development theory
has reached some kind of impasse because it has been

unable to explain the growing number of instances where several different models co-exist and overlap in the same country. Instead of recognizing the merits and defects of all positions and trying to test their application on the basis of the concept of historical specificity, development theory tends to be fixed in typologies of universalism.

Paulette Pierce's study of Guyana from independence to the present is a valuable, theoretically informed case study in third world development that clearly demonstrates that the paths to genuine autonomy do not lead from independence in any kind of causal nexus. Her main arguments, (1) that liberal democratic state forms inherited from western powers are compatible with dependent relations and that (2) marxist-leninist ideology and nationalization are compatible with capitalism, will make none of the proponents of modernist, leninist or dependency versions of post-war third world development happy. Yet, this carefully researched and detailed account of the course of Guyanese development since independence is a highly persuasive counterpoint to all of the received wisdom.

There is no question that a quarter of a century of independence has left Guyana in a state of economic dependency, one-sided economic development and persistent political and social turmoil. As Pierce shows, its class relations are severly mediated by ethnic and racial conflicts between its two large groups, the East Indians and the Blacks. Her subtle and nuanced account of this core of the political struggle within the country is important not only for those interested in the relationship between race and class, but also for its many side analysis of the vagaries of ideological conflict and cooptation. The Guyanese adopted a modified verson of the British paraliamentary system, like many other former colonies such as India, Canada and Australia. Both leading parties proclaimed themselves to be socialists, although the East Indian dominated Peoples Progressive Party (PPP) adopted a marxist-leninist ideology which identified nationalization with the transition to socialism. PPP's influence over the large sugar workers union and its considerable voting strength helped drive its more moderate rival, the PNC, led by Prime Minister Forbes Burnham, to the left. In the past fifteen years, Burnham has taken over a large part of the PPP program--nationalization, marxist-leninist ideology, and other features--but, as Pierce shows, was never far from its dependency relationship with foreign capital, the International Monetary Fund and the World Bank. In Guyana, the prerequisite for modernization was present--parliamentary

democracy; the leninist prerequisite for independent
economic development--the abolition of private property in
the largest enterprises, especially sugar; and the rhetorical socialist ideology and welfare program. However,
Pierce's critical and unsentimental study comprehends that
these forms do not obviate the fact that the state acted
as a capitalist, that democracy and political repression
seemed to go in tandem and that marxism could as easily be
understood as a modernization ideology within a state
capitalist framework as a revolutionary doctrine. These
juxtapositions are anamolous only if one remains ensconced
in the old antimonies. The force of Pierce's analysis is
to delineate the concrete conditions that made all of them
possible and relatively coherent. One of the most fascinating and telling themes in the book is the description
of the complex relations between Guyana's largest foreign
company, Bookers, the country's major sugar producer, the
state and the political and labor structures. Bookers was
more than a paternalistic employer; Pierce shows how much
it devised a new corporate liberal model of colonial
power. Here we have a foreign company that was prepared
to institute affirmative action to replace its old white
managerial cadre with East Indians and Blacks. Moreover,
it prepared itself for nationalization when it became
evident, in the course of the explosion and then collapse
of world sugar prices in the 1970s, that the Burnham
government had no choice but to take over its facilities.
Yet, the company retained its close working relationship
with the government and assisted Burnham to achieve and
then retain political power. Pierce describes the degree
to which the old colonial model of brutal repression was
replaced by a neo-colonial system in which political
independence and state ownership of key industries was
initiated without changing the basic relations of economic
power that were established under the old regime. How-
ver, "brutal" repression, although no longer the norm,
could still be selectively employed to deal with those
like the insurgent Working Peoples Alliance (WPA) that
refused to play by the rules of the now routine political
game to which the opposition had more or less comfortably
adjusted. Moreover, the government did not hesitate to
establish its own wholly dominated institutions such as
company trade unions to foil the efforts of the opposition
to work through the labor movement to achieve political
power.

The picture portrayed in this work is of a complex,
multilayered society in which political, economic and
social relations defy conventional categories. As Pierce
shows, with dependency theorists, the lack of a powerful
local bourgeoisie does not signify a post-capitalist,

post-colonial society since the state can assume these functions. But she departs from garden variety dependency expectations when she shows that political parties and the state are capable of wide-ranging ideological and policy changes that, for a time, thwart the emergence of a genuine opposition because these forms are flexible enough to override the conservative influence of bureaucracy and tradition. In these sections, Pierce confounds the certainties of marxist writers for whom the leninist creed is the exclusive property of the intellectuals who organize the revolutionary party and its working class and peasant followers. Burnham, whose party broke with the once dominant PPP, proved a formidable and sophisticated leader when he was able to use marxism-leninism as a strategy for mass mobilization against both his left and right wing opponents. In a country where all three socialist political organizations held the same ideology, the uses of doctrine are confusing to the dogmatist. Pierce's sympathies are with the WPA, whose base is still fairly narrow, having won mass support only among bauxite miners. She urges powerfully that its source of potential strength in contrast to the two leading parties, is that it is multiracial and multiethnic. That is, according to Pierce, marxism cannot justify itself except as a class theory and strategy and she finds the WPA trying to move in that direction.

I have no doubt that Pierce's book, if read dispassionately, will change development studies. At the very least, it should constitute a pioneering approach to third world studies, upon which others can build. Although she has not fully drawn the theoretical implications of her study, I believe she has demonstrated a new model for understanding the new colonialism. This book will be around for a long time.

Stanley Aronowitz

June 1984

Noncapitalist Development

Chapter 1

Overview of the Problem and Theoretical Issues

In May 1976, Guyana celebrated its tenth anniversary of independence by nationalizing the property of Booker McConnell, Ltd., a British expatriate firm which owned the country's sugar industry. This dramatic event symbolized the culmination of the radicalization of the development strategy of the Black ruling party. In 1970, the Peoples' National Congress resurrected its socialist image and declared Guyana the world's first Co-operative Socialist Republic. In 1971, the Government nationalized the majority of the country's bauxite mining industry. In 1975, it took over the remainder. Thus sugar, the first major impetus to Guyana's colonial exploitation, stood out more sharply than ever as a challenge to the PNC's commitment to implement its policy of complete ownership and control of local resources. This was the last major step in the removal of the symbols of foreign exploitation, and it was the hardest to take. On the one hand, sugar workers are the most exploited segment of the Guyanese work force and personify the peoples' history of enslavement, indentureship and super-exploitation as wage laborers. If the Government really meant to lead the nation in a socialist revolution, this glaring reminder of expatriate control would have to go. On the other hand, sugar workers are overwhelmingly East Indian and loyal supporters of the Peoples' Progressive Party, the largest opposition party in the country. To take sugar the PNC had to be willing to walk into the lion's den.

Guyana, like most postcolonial societies, has a history of deep, racial divisions. Within its small population of under a million people, six racial groups--East Indian, Black, Chinese, Portuguese, Amerindian and "colored"--are represented. Their respective positions in the color-class hierarchy reflect the circumstances of their introduction into the territory by the colonial power. Blacks were brought in as slaves to work in the sugar fields and primitive cane grinding factories. After the abolition of slavery, the planters first tried Chinese and Portuguese workers as substitutes and finally settled upon the large-scale importation of indentured East Indian labor. Blacks subsequently migrated to the urban areas and provided a source of cheap, unskilled labor. The Chinese and Portuguese moved into the retail business with the help of the British colonizers. The Amerindians, the indigenous population once used as slave catchers, remained in the hinterland completely outside of the colonial class structure. In this artificially created

society, racial suspicions and hostilities constantly
brewed and, when it was to their advantage, the British
would stir the pot. Thus when the Peoples' Progressive
Party, the first, mass-based independence party was
formed, the British applied pressure to its weakest
link. Submerged conflicts within its East Indian and
Black leadership were expertly manipulated and the party
shortly split in two. Cheddi Jagan, an East Indian and a
marxist-leninist, maintained the support of the East
Indian population--particularly those who were members of
GAWU, the powerful union in the sugar industry. Forbes
Burnham, a temporizing socialist and Black leader, drew
the support of the Black population and formed a rival
party, the Peoples' National Congress. In order to gain
political power, Burnham moved his new party steadily to
the right until he eventually formed an alliance with the
United Force, an ultra-right wing party representing
bourgeois interests in general and the Portuguese in
particular.

With this bitter legacy of racial and class conflict
in mind, the PNC's recent Marxist-Leninist conversion and
decision to nationalize sugar calls for careful analysis.
Indeed it provides the material for a case study of the
struggle to achieve socialism in a racially torn and
underdeveloped society. As we will discuss below, the
role of racial and class conflict in social change in
Third World countries has bedeviled mainstream and
Marxist scholars alike.

Since the end of World War II, but to a lesser extent
today, mainstream social science has tried to explain the
patterns of social structure and change characteristic of
"developing societies." The most recognized and controversial achievement in this regard for the Carribbean
and Africa has been the model of a "plural society" which
provided the theoretical framework for much of the anthro-
pological and sociological research during the 1960s. The
concept was first developed by a British colonial officer
reflecting upon his experience in the Far East.

> . . . In Burma, as in Java, probably the first
> thing that strikes the visitor is the medley of
> peoples--European, Chinese, Indian, and native.
> It is in the strictest sense a medley, for they
> mix but do not combine. Each group holds by its
> own religion, its own culture and language, its
> own ideas and ways. As individuals they meet,
> but only in the market-place, in buying and
> selling. There is a plural <u>society</u>, with
> different sections, of the community

but separately, within the same political unit.
Even in the economic sphere, there is a division
of labor along racial lines.[1]

Furnivall's ideas were picked up by many scholars, particularly those like M.G. Smith, Leo Despres, Harry Hoetnik,
Neville Layne and others who were interested in the
Caribbean.[2]   According to these theorists and research-
ers, the key determinant of social structure and social
change in West Indian societies is the existence of
racially and/or culturally distinct and competing segments
within the population. Societal integration and social
peace is only maintained to the extent that one segment,
previously the white colonial group, has a monopoly of
coercive force and economic sanctions. The introduction
of significant social or economic changes will, the plural
model predicts, unleash the forces of pluralism and lead
to inter-communal conflict often to the point of blood-
shed.
     The picture presented is grim and, we suspect, subtly
racist. Without the supervision of a benign colonial
master, it was implied that the local populations would
inevitably dissipate their energies in primordial warfare.
Therefore, after two years of racial bloodletting and
imperial intervention, Guyana was held up as an ideal-
typical plural society. Conspicuously absent from these
interpretations of the Guyanese and other postcolonial
situations were (1) the determinative role of economic and
political forces based in the metropolitan countries and
(2) the nature and dynamic of the local class structure.
Following independence, the activities of imperial forces
mysteriously vanished from the theoretical consideration
of those utilizing the plural model. Consideration of
stratification within these ethnic or racial groups was
minimal so that the plural segments could be depicted as
monolithic blocs.
     Starting in the late sixties however, adherents of
the plural model began to modify their position to acknowledge the increasingly apparent elite group manipulation of ethnic sentiments. Backing away from Clifford
Greetz's assumption of the inevitability of primordial
conflict, scholars such as Crawford Young began to talk
about the situational character of ethnic identification
and mobilization. The theoretical development did not go
far enough; no serious attempt was made to account for the
role of external and internal class-rooted factors. The
short-circuited character of this revised approach is
exemplified in the conclusion reached by Katherine West,
". . . whatever their origins, once aroused, ethnic

hostilities assume a life of their own . . . ."[3] Leo Despres, an anthropologist who has done extensive research on Guyana utilizing the plural model, arrives at a similar conclusion. Although he recognizes the impact of international political forces, he nonetheless does not integrate these observations into his theoretical considerations.

It is precisely this failure of the plural model to address the external political environment and the local class structure which troubles R. T. Smith in his critique of its application to Guyana. No one familiar with the Guyanese situation would try to discount the significance of racial and cultural symbols, but, Despres' attempt to reduce the country's political struggles to simple, racial chauvinism is, in Smith's opinion, "bizarre."

> If the basis of political conflict in Guyana is simple racial and cultural communalism, then the ideological posture of political parties and leaders would be of little account since it would simply be a mask behind which hides the real face of ethnic sectionalism. Yet we know that ideological posture has been, and presumably still is, of the greatest importance to those who decide which Caribbean governments are "acceptable" to the interests of the United States of America.[4]

Furthermore, Smith notes that it is curious that racial sentiments were strongest in the culturally homogenized middle class communities of Georgetown since, according to the plural model, ethnic identification is supposed to be strongest within groups most dissimilar. With decolonization on the horizon, Guyana's middle class factions did grasp at the symbols of racial solidarity in their competition for upward mobility but this was not the whole story. The two major political leaders to emerge out of the independence struggle came to represent antagonistic class interests. Thus when the communist Jagan and the moderate Burnham met with Kennedy and Schlesinger in Washington during 1961 the latter was chosen. Such external manipulation of the course of social development is not, Smith concludes, unusual in "these poor, small, and weak ex-colonial territories"[5] and our theoretical models must take this into account.

On the other side of the ideological spectrum, marxists have almost exclusively focused upon external factors, such as how the operation of world capitalist markets and the activities of imperialist states ensure the reproduction of underdevelopment in the Third World. It is only in recent years that marxists have undertaken the difficult task of working out a theory of under-development based on an analysis of the local class structure, the interpenetration of class and racial divisions and the role of the local state in social development. In large measure the delay was due to the enormous popularity enjoyed by dependency theory through-out the sixties and the first half of the seventies. According to Andre Gunder Frank, the founder of the quasi-marxist dependency school, colonial analysis which "cen-ters on and emerges from the metropolis-satellite structure of the capitalist system" is the foundation of dependency theory.[6] Class analysis which in traditional marxist methodology centers on the examination of the national class structure determined by the mode of pro-duction and the role of the state in maintaining a par-ticular system of exploitation receives only perfunctory mention.

The basic tenets of dependency theory can be summar-ized as follows: (1) The world capitalist system is composed of two opposing parts--the wealthy core (metrop-olis) and the poor, exploited periphery (satellites). (2) The core and the periphery are inseparably linked and therefore can only be understood in terms of their rela-tion to each other. (3) Capitalist expansion is in-herently contradictory, simultaneously creating develop-ment in the core and underdevelopment in the periphery. (4) The principal contradiction within the world capital-ist system is no longer located within the advanced industrial nations. Today, the fate of capitalism depends upon the outcome of struggles between imperialism and anti-imperialist forces located in underdeveloped coun-tries (UDCs). (5) And finally, revolution and the des-truction of the world capitalist system are the only route to development for the countries of the Third World.[7]

Clearly, dependency theorists see the capitalist system as based upon exploitation. Its departure from more traditional marxist analysis is due to the unusual manner in which this exploitative relationship is con-ceptualized. For Andre Gunder Frank, regions and entire nations take the place of social classes in explaining the process of capital accumulation. The metropolis exploits its satellites and the core regions within underdeveloped countries exploit their own backward regions.

> This contradictory metropolitan center-peripheral satellite relationship, like the process of surplus expropriation/appropriation, runs through

> the entire world capitalist system in chain-like
> fashion from its uppermost metropolitan world
> center, through each of the various national,
> regional, local, and enterprise centers.[8]

The inevitable consequence of this geographically conceived pattern of exploitation is deepening underdevelopment and increasing dependency for the periphery.

> Thus the metropolis expropriates economic surplus
> from its satellites and appropriates it for its
> own economic development. The satellites remain
> underdeveloped for lack of access to their own
> surplus and as a consequence of the same polar-
> ization and exploitative contradictions which the
> metropolis introduces and maintains in the
> satellites' domestic economic structure.[9]

Unfortunately, Frank fails to analyze this internal
structure which he notes "the metropolis introduces
and maintains" in order to ensure imperialist control of
the global process of capital accumulation. Instead, he
and other dependency theorists have concentrated on
external factors--the international division of labor, the
world market, private foreign investments, the IMF, aid
programs and private financial markets. Much of the work
has been excellent.[10] The continuous drain of surplus
from the impoverished periphery to the increasingly
wealthy core nations has been revealed in broad outline.
However, the ways in which this system of exploitation is
changing and how it is to be overcome are not addressed.
This, as Frank is well aware, would require a detailed
analysis of the development of Third World economies and
their class structures. Because he instead chose to
concentrate on the colonial structure his essays ". . .
cannot, nor are they intended to, serve as an adequate
instrument to examine the class struggle as a whole and to
devise the necessary popular strategy and tactics for its
development, for the destruction of the capitalist system,
and thereby for the development of the underdeveloped
countries."[11]

From the start, dependency theory aroused criticism
among those who objected to its unorthodox approach to the
issue of exploitation. Cabral Bowling and a group of
Mexican critics challenged Frank's analysis precisely on
this point: "We believe that exploitation is a social
phenomenon too complex to be explained exclusively in
terms of the metropolis-satellite structure. . . . Would
it not be more accurate to state the relations of

exploitation in terms of social classes?"[12]   To this, Frank replied yes and that his work was indeed intended as a guide to such an analysis. In response, Theotonio dos Santos denied the compatibility of the two modes of analysis and more importantly concluded that Frank's methodology is at bottom static.

> The colonial pattern which Frank outlines cannot be combined with class analysis, as he seeks to do . . . As for changes in the system, it is not enough to demonstrate the persistence of the colonial structure. It is necessary to explain how the forms of dependence have changed, in spite of its persistence. For these changes have produced the profound contemporary crisis, which both requires and facilitates a socialist solution.[13]

Despite these early and perceptive criticisms and the resulting efforts to infuse dependency theory with class analysis, little substantial progress was made until the late 1970s.[14]

An essay by James Petras and Kent Trachte entitled "Liberal, Structural, and Radical Approaches to Political Economy: An Assessment and Alternative" is probably the best attempt to date to integrate a world system perspective with a more traditional marxist approach to the problem of exploitation.[15]   According to Petras and Trachte, dependency theory and the more recent world system approaches make the fatal mistake of defining the problem as dependency versus exploitation. Moreover, by selecting the world capitalist system as the primary unit of analysis and using analytic concepts "which obscure the real historical actors," those approaches make it impossible to understand the historical transformation of that system.

Dependency and world system theorists have turned the problem on its head:

> It is not the world system that begets change in social relations, but rather social forces that emerge and extend their activities that produce the world market. The transformations wrought within societies by their insertion in the world market   must be seen as an ongoing reciprocal relationship: between the forces and relations of production within a social formation and those that operate through the world market. From the perspective of international political economy, a

> comprehensive analytical framework must focus on
> the structural variations and transformations
> within the capitalist mode of production.[16]

The capitalist mode of production is first and foremost
a system of labor exploitation based on the extraction
of surplus value. Therefore, Petras and Trachte argue
that we must focus on the creation, reproduction and
expansion of the capital relation within both the core
capitalist countries and the dependent satellites rather
than concentrate on the abstract external mechanisms
which link the two. We should therefore concentrate on
how as capitalist production expands within a particular
country it transforms the local class structure, the
potential for different forms of social conflict and the
structure of the state. As the strength, position, and
number of classes change, old alliances will be modified
or completely abandoned. All of this goes on within the
larger context of the world market; but, the market is not
the fetish that early dependency theorists protrayed it to
be. "The world market operates through the class-directed
institutions that impose the exploitative class relationship throughout the world."[17]

Consequently, the class structure and the class
character of the state in underdeveloped countries (UDCs)
are not a secondary concern as implied by dependency
theory.

> In understanding the processes of world historic
> change it is not as important to know that Zaire
> and Mozambique are peripheral societies as it is
> to recognize the profound class transformations
> that affect one and not the other . . . their
> participation in the world capitalist market is
> informed by a different set of class interests
> which act decisively to effect the character and
> shape of the development of the productive forces
> within society and to contribute to undermining
> the organizing principles of the world capitalist
> system in the larger historical perspective.[18]

Therefore if core capital is to retain its control of
the global accumulation process, imperialist states must
prevent the emergence of class alliances and state policy
in UDCs which would block access to the surplus value
produced in the periphery. A static conception of metropolis-satellite exploitation and an analysis of unequal
exchange, cannot explain this ongoing struggle: "Insofar
as imperialist forces act, they operate within the class

formation and cannot be conceptualized as the impersonal forces of the market . . . they must be marked as a part of the internal class alignments."[19] The process is further complicated by the dynamic character of the internal class alignments in both core and peripheral societies. The structure of imperialism and underdevelopment is therefore constantly changing as the dominant segments of capital in the advanced countries seeks to respond to domestic and foreign crises and national capital within UDCs struggle to prevent its complete subordination and absorption.

The state plays a crucial role in the process of capital accumulation. In advanced capitalist societies the state both reflects and shapes the class structure as it acts to ensure the fundamental capital/labor relation and economic expansion. However, in UDCs the principal determinant of the structure and development of the state is not the imperatives of local accumulation but its relationship to the imperialist states. For Petras and Trachte this imperialist imperative to control state policy in the periphery is self-evident. "The state is the critical unit in the process of converting class alliances into development strategies . . . [it is] . . . the crucial instrumentality in this process of reversing regimes, reconcentrating income, and reopening economic channels . . . ."[20]

Prior to decolonization, the exploitation of the Third World was "relatively direct." The colonial classes were directly exploited by foreign capitalists. The local state was merely an extension of the imperial state. However since independence, the process of exploitation has been complicated by the rapid growth of a local intermediary stratum which for the most part is propertyless and wholly dependent on the state for its political power and the economic means to consolidate itself as the new ruling class. This group now stands between imperial capital and the local labor force creating new contradictions and disguising old ones.[21]

The key to Petras and Trachte's conception of the process of underdevelopment is the idea that peripheral exploitation can be maintained by at least three types of class alliances and their corresponding development strategies. In the neocolonial model the national bourgeoisie, the petty bourgeois state and imperial capital form an alliance to increase the exploitation of local labor. The state's development plan includes handsome investment incentives, and the promise of a large, disciplined and cheap labor force. "Whatever the specifics, the foreign component is clearly dominant in internal, as well as external, relations."[22]

In the second strategy, called the <u>national bourgeois developmental approach,</u> the state seeks to gain control of local accumulation in the interests of national capital and/or the petty bourgeois state. Under a nationalist or socialist cover, the state moves against foreign capital. The new development policy may include sharp tax increases on foreign firms, restrictions on profit repatriation or investment and even nationalization. Nonetheless, the anti-imperialism of the regime does not extend to include an alliance with the working classes, "while squeezing the foreign sector, [it] also shares with the foreign sector an interest in maximizing exploitation of the labor force: maintenance of production, labor discipline, and popular demobilization."[23]

The third strategy, the <u>national-populist alliance,</u> is a frequent occurrence but the least stable of the three. Herein the the bourgeoisie and the petty bourgeoisie join with the working class and peasantry in an anti-imperialist alliance. At first there is general agreement that foreign exploitation should be eliminated but, the alliance quickly falls apart over the question of which social class should now control the accumulation process. Without its foreign allies, the bourgeoisie has no alternative but to push for expansion at the expense of the working class and peasantry. The poor naturally resists any attempt to freeze or reduce their standard of living. If the state sides with the bourgeoisie, the alliance turns into a national developmental approach which is characterized by extensive repression--as is the neocolonial model. And in all probability, this alliance will again be changed to include foreign capital once the local bourgeoisie demonstrates its inability to develop the economy. If, however, the state takes the less likely course and moves further to the left by attacking local as well as foreign capital, it "alienates the bourgeoisie and leads to an alternative non-capitalist model of capital accumulation."[24]

The three types of alliances and the resulting patterns of exploitation are not mutually exclusive. And, as Petras and Trachte are well aware, any one "is difficult to identify as a pure type."[25] In fact, we do no damage to their analysis by thinking of the alliances as recurring phases in the process of capitalist exploitation of the periphery. Needless to say, "imperialist relations are central in sustaining or destabilizing the two polar types of social regimes."[26] In the case of the neocolonial alliance, the imperialist state engages in "state-building" functions. For example, local bureaucrats and especially members of the military and the police receive Western training.   Western experts are on hand to advise the local political elite.   Friendly governments receive generous grants and aid and lines of credit.   Should however the government turn towards a national development approach or a national popular alliance, the imperialist state moves quickly to "disaggregate" the state utilizing local social and political forces or external mechanisms such as military intervention or a credit squeeze.   Through these two processes, state-formation or disaggregation, imperialism imposes its exploitative class relations on the periphery.   But as Petras' and Trachte's analysis makes clear, it is not the formal structure or rhetoric of the postcolonial state which is of concern but rather "access to the internally generated surplus and the creation of class relations which facilitate access."[27]

Can a popular alliance of all nationalist forces led by the petty bourgeoisie bring an end to peripheral exploitation?   Petras and Trachte are not optimistic. Although capitalist expansion leads to the development of class forces favoring a leftist alliance and a noncapitalist model of capital accumulation, the imperialist state works to abort this kind of revolutionary development. Reflecting on Petras' and Trachte's analysis we are left with a gloomy impression of the postcolonial state vacillating between neocolonial and popularist alliances with the center of gravity favoring the neocolonial model.

Theories of Noncapitalist Development

Conversely, the proponents of the theory of noncapitalist development are quite sanguine.[28]   Whereas Petras and Trachte stress imperialist domination of the state in Third World societies and the susceptibility of petty bourgeois leadership to counter-revolutionary attack, the theory of noncapitalist development draws attention to the unusually high degree of autonomy enjoyed by the post-colonial state and hence the revolutionary potential of a petty bourgeois leadership armed with state power.   The notion of a third path of social development-- neither capitalist nor socialist--was first introduced by Lenin in 1920 but it was not further developed until the global anti-colonial movement was in full swing.   In 1960 at a world conference of Communist and Workers Parties convened in Moscow, Soviet scholars put forth an elaborated thesis of noncapitalist development.   By the late 1970s it was the center of a hot debate which raged at a conference on socialism and developing countries held in

Yugoslavia.   The theory has held particular fascination
for Caribbean politicians and leftist scholars.

John Ohiorhenuan summaries the position as follows:
"Under certain circumstances, the post-colonial state, led
by an alliance of progressive forces and supported fully
by the socialist countries, can fulfill the 'historic'
role of a capitalist class (vis., accumulation), and in so
doing, can create the conditions for a socialist transition."[29]   The Soviet message was full of hope.   Without
armed struggle or developed industrial capacity, backward
countries could begin what is called "the transition to
the transition to socialism."   Between national independence and the creation of a socialist society there stands
the necessary phase of noncapitalist development or
national democracy.   This phase can be readily reached by
declaring an anti-imperialist, anti-capitalist and anti-
feudal policy.

According to the theory of noncapitalist development
such daring policy reversals are possible because the
postcolonial state is relatively free from the usual class
constraints which limit the action of the state in advanced capitalist societies.   This is part of the janus-
faced legacy of colonialism.   Because of underdevelopment
there are no powerful domestic classes with firm roots in
the productive forces of the society capable of dictating
state policy.   The peasantry is generally large but
unorganized and apolitical.   The working class is small
and fragmented.   The state therefore emerges as the decisive force in social development once the control of the
exploiting colonial classes is shaken by independence.   In
the absence of the type of class politics characteristic
of the developed countries, the state can form a national
patriotic alliance to defeat imperialism, develop the
economy and mobilize the masses.   In short, the state can
be the principal institution for constructing a socialist
society.   This needless to say, is a fascinating reversal
of the traditional marxian view which holds that the state
is the quintessential expression of capitalist exploitation which must be smashed by the revolutionary forces.

The challenge to traditional marxism has not gone
unnoticed.   Jay Mandle points out that in Jamaica (under
Michael Manley) and Guyana the same governments which
pursued strictly capitalist development plans abandoned
these and adopted socialist development strategies.   This,
in Mandle's view, throws into question the idea that the
structure and personnel of the state must be radically
changed if not completely abolished in the process of
socialist revolution.   Mandle concludes that in the case
of Jamaica and Guyana "the shift to the noncapitalist path

was essentially an administrative action." This does not trouble him. For Mandle, "essentially, the question is whether this process can be reversed and whether the non-capitalist path will result in a social organization in which political leaders are responsive to the population generally and the working class in particular rather than vice versa. In short, what are the prospects that the non-capitalist path will lead to socialism?"[30]

Not surprisingly, this assumption that revolutionary change can be instituted by administrative directive has disturbed many involved in the struggle for socialist transformation. According to Charles Bettelheim, a revolutionary party must refrain from imposing orders on the masses because this will eventually lead to authoritarian rule, not to socialism. "A proletarian party cannot claim to 'act in place of' the masses. For the masses must transform themselves while transforming the objective world and they can transform themselves only through their own experience of victories and defeats. This is the only way in which the masses can achieve a collective consciousness, a collective will, and a collective capacity, i.e., their freedom as a class."[31] Ohiorhenuan reflects this same discomfort when he warns that ". . . the non-capitalist thesis has down-played the necessity for the independent development of the working class movements . . . greater attention must be paid not only to the origins of any government declaring itself socialist, but also to the conditions of the working class and its relative strength in the alliance."[32] In the context of post-colonial societies, many have taken these conclusions one step further and argued that the existing form of the state inherited from the colonial ruler must be completely abolished.

Its structures were designed to maintain colonial dependency and they cannot now miraculously spearhead the movement for independent economic development. Amilcar Cabral personified this view:

> We are not interested in the preservation of any of the structures of the colonial state. It is our opinion that it is necessary to totally destroy, to break, to reduce to ash all aspects of the colonial state in our country in order to make everything possible for our people. The masses realize that this is true, in order to convince everyone that we are really finished with colonial domination in our country.[33]

The petty bourgeois leadership of the masses must, Cabral insisted, be ready to commit class suicide. They must cease to represent the interests of the petty bourgeoisie and completely identify with the working classes. This is not simple. It ordinarily takes very difficult circumstances to foster this type of unity. Without romanticizing the role of violent struggle, we can appreciate the unique opportunities it presents to the political elite and the masses in the Third World. In the case of the former, it forces a petty bourgeois leadership into a dependent relationship with the working class and peasants which social position and cultural breeding have previously separated them from. In the case of the latter, it forces the masses to organize and make their own decisions outside the boundaries of traditional or imperial authority. As Harry Magdoff observes, violent struggle does not guarantee that the transition to socialism will be successful but ". . . where the colonial or neocolonial state has been smashed and . . . there has truly been a transfer of power to the most oppressed, and if the masses have been mobilized to exercise this power,· there is a good chance that it will."34

In his critique of non-capitalist development theory, Harry Magdoff argues that most of the recent attempts by Third World governments to radicalize their development strategies will fail. This is because they are trying to develop the economy within the context of the existing class structure. A few bureaucratic changes and the pronouncement of a socialist ideology will not do the trick. The oppressive and exploitative class relationships of the country will continue to choke off productivity and the chronic foreign exchange crises will force the rebellious governments back into the imperialist fold. Despite the promises of African, Arab and Co-operative Socialism, "the contradictions and constraints which characterize this kind of society are such that the result is likely to be not a noncapitalist road to socialism, but a new variant of capitalism that in the final analysis remains dependent on the imperialist centers."35

Theories of Bourgeois Revolution and State Capitalist Development

Ian Roxborough's perspective on the recent trend of political and economic radicalization in the Third World is also opposed to that of the theory of noncapitalist development. Roxborough strongly objects to viewing these developments as the precursors to a socialist revolution

or as the "transition to the transition." On the contrary, the aggressive policies of Third World states are,
in Roxborough's opinion, a new phase in the bourgeois
revolution which has never been completely accomplished in
the periphery. The inability to accomplish the historic
task of the bourgeois revolution--a viable nation-state
and autonomous capital accumulation--is what characterizes
and perpetuates underdevelopment. Whereas this transition
to the capitalist mode of production occurred in an
"abbreviated space of historical time" in the West it has
spread over centuries in the underdeveloped world.
Moreover, the economic and political achievements have
been provisory and "constantly subject to interruption and
historical retrogression." New forms of dependency and
exploitation are always on the horizon.[36]

As a consequence of the uneven nature of capitalist
development in the periphery an "exceptional" state form
has developed in most Third World countries. The pre-
capitalist social formation has been undermined, but, the
two key classes of capitalist society (the bourgeoisie and
the working class) remain frozen in an embryonic stage.
This distorted pattern of development has ". . . enabled
the elite which come to occupy state power to transform
themselves into new dominant classes."[37] The most
recent spate of attacks on foreign capital is simply the
latest in a series of efforts to complete a bourgeois
revolution in the periphery. The confrontation with
capital can often become quite intense and vitriolic.
This fact, however, and the tendency of the political and
bureaucratic elite to defend their actions in terms of
marxism-leninism does not, in Roxborough's opinion, make
for a bona fide socialist revolution. The anti-
imperialism of the regimes and the vastly expanded role of
the state in the economy are strategies aimed at winning a
better position within the world capitalist system.
Furthermore as with previous attempts to consolidate an
independent bourgeois class in the periphery, it will,
Roxborough predicts, most likely fail and degenerate into
a form of neo-dependency. This is not said to preclude
the possibly progressive and/or popular character of these
developments. Sometimes they are both. Roxborough
does, however, insist that we use the concepts of social-
i·sm and proletarian revolution with greater precision.

> In many current versions of Marxism, "proletarian
> revolution" has lost its original meaning of a
> revolution carried out by the proletariat to
> establish socialism, and has come to mean simply
> a process which results in the creation of a

> state committed to some form of economic plan-
> ning, state ownership and economic growth. When
> these are features of nearly all forms of con-
> temporary economic systems it is hardly sur-
> prising that "socialism" is so widespread. The
> class nature of these regimes needs, however, to
> be examined with more care . . . .[38]

This confusion of state ownership, centralization,
national planning and growth with socialism stems from the
transmutation of marxism in historical practice. During
the consolidation of the Russian Revolution, marxism
ceased to be the revolutionary theory of the industrial
working class for its self-liberation against the forces
of capital. In Stalin's hands, it became the recipe for
rapid economic growth through the liberation of the forces
of production and the ideological justification for an
emerging dominant stratum Djilas has dubbed the "new
class." In complete contradiction to their original
intent, marxian principles were interpreted to prescribe
greater labor discipline imposed by the party and the
labor bureaucracy under its control. Mao's subsequent
interpretations of marxism completed the dissociation of
Marxism from an analysis based on the social structure of
advanced industrialized societies. According to Mao, the
peasants of the underdeveloped world would be the grave-
diggers of capitalism. From bitter experience he had lost
his faith in the revolutionary potential of the working
class. The ambiguous role of the working class in the
Cuban revolution and vanguard position of the petty
bourgeoisie, pushed the original meaning of marxism
further into the background of practical revolutionary theory. "The word [proletariat] changed its meaning;
it no longer referred in any way to a special social
class; rather it identified a particular constellation of
ideological themes."[39]

Consequently, in countries where the industrial
working class is virtually nonexistent or where it is
small and/or divided we hear claims of the victory of the
socialist revolution. Roxborough relying upon an orthodox
interpretation of marxist theory, argues that this cannot
be the case. The challenge is therefore to specify the
class character of these regimes which claim to be socialist. Roxborough's assessment is extremely alarming:

> The socialism which exists in many countries of
> the Third World is indeed a lumpen-socialism.
> What exists is a form of class rule in which the
> historical task of capital accumulation

> (abdicated by the bourgeoisie) is performed by a bureaucratic elite drawn from diverse petty bourgeois sectors. This elite retains state power through the most varied forms of corruption, nepotism and repression while it attempts to consolidate itself into a new capitalist class . . . Thus it is that a lumpentheory of lumpendevelopment produces in its turn a lumpensocialism. The confusions stem from the analysis of social classes from the assimilation of the social relations internal to the social formations of dependent societies to the model of colonial relations.[40]

Thus, according to Roxborough, the best which could emerge from the recent trend of radicalization in the Third World would be the success of a bourgeois revolution and the consolidation of a new capitalist class dominant in the domestic economy and unquestionably in control of the national state. The worst which could emerge would be a perverted form of socialist development leading to new forms of dependency on the imperialist centers and the repression and demoralization of the local population.

James Petras in an article entitled "State Capitalism and the World Third" draws conclusions similar to Roxborough.[41]   State Capitalism is the unique development strategy of the "new intermediary stratum" (state sector employees) in the Third World today. It represents the final attempt at national capitalist development in the periphery--"the nationalist state remains the last barrier to total subordination and fragmentation in the new international division of labor."[42]   In the previously discussed article written with Trachte, Petras identified three strategies of capitalist accumulation in the periphery (the neo-colonial, national bourgeoisie developmental approach and the popular alliance), now he identifies what could best be considered a fourth. When both the national bourgeosie and the popular classes are too weak to form the ruling alliance and the neocolonial model has failed to lead to an integrated industrial development, the bureaucratic elite will strike out on its own to capitalize the national economy. In Petras' words "it is possible to conceive of a class-conscious stratum vertically and horizontally linked functioning as an independent class (apart from workers and bourgeois) with its own political-economic project."[43]

Although this state sector elite adopts socialist forms, i.e., marxist-leninist ideology, the one party state, mass organizations, state ownership and centralized

planning, its objective is to tie "the expansion of
capitalist market relations to the expansion of the
state."[44] The goal of research should be to understand
how the social structural position of this new stratum
shapes its vision of the new social order. For instance,
Petras points out the close relationship between the
bureaucratic milieu of this ruling stratum and its fond-
ness for state capitalist solutions to the crisis of
accumulation. "Hierarchical order and functional inter-
dependence, the two characteristics of the bureaucratic
order, lend themselves to a political ideology that
embraces collectivism without redistribution."[45]

The next question pertains to the international
context which facilitates the emergence of the state
capitalist strategy of accumulation. We have already
noted the weaknesses within the internal class structure
which allows the intermediary stratum to function as an
independent class but this alone does not lead to the
rejection of the neocolonial path of development. The
state elite has to be able to see openings within the
world capitalist system, that is, the chance to successfully renegotiate the existing pattern of exploitative
relations set in place in the post-W.W. II period. The
increasingly evident decline in the hegemony of the United
States has, Petras believes, signalled this opportunity.
The decline in American power and the resulting increase
in inter-imperialist rivalry has meant that Third World
states can buck the usual terms of neocolonial alliance
and escape or minimize the most severe forms of U.S.
retaliation, i.e., economic blockade, military inva-
sion, or a credit squeeze. To the extent that the U.S.
can no longer dictate the foreign policy of other advanced
capitalist countries, there are alternative sources of
trade, capital, technology and aid available to the
rebellious state in the periphery.

Nonetheless, Petras does not believe that the state
capitalist strategy of accumulation will succeed in
developing the economy or enable the state to legitimate
its rule. Instead the regime will become more and more
isolated. Its strategy of alternately placating one class
while attacking another eventually loses its effectiveness
as it becomes increasingly clear to the national bourgeoisie and the working classes that the bureaucracy seeks
to "impose its own imprint on society." Mutual suspicion
and hostility between the national bourgeoisie and the
state will grow more intense despite the fact that the
private sector is rarely eliminated as such. The efforts
on the part of the working classes to increase their share
of the surplus will be ruthlessly crushed. The

opportunities afforded by inter-imperialist rivalry are never as great as are hoped for. "The effort to create a political base independent of imperialism on the one hand, and 'above' the working/peasant masses on the other, leads to a vunerable political situation, one in which divisions within the junta can easily lead to reversion to a pro-imperial regime."[46]

Unlike Roxborough, Petras does not foresee the possible consolidation of a new ruling class. This presupposes a process extending over decades and leading to the development of a new productive forces and rela-tions. In state capitalist society "the sharp contra-dictions emerge from the very inception of the regime"[47] and shortly threaten the government's continued capacity to rule. In response the state bourgeoisie will often-times turn to racial and ethnic ploys to mobilize a constituency which transcends class boundaries to defend its hegemony. For the victimized ethnic/racial group the country can become a "prison."[48] Interestingly, enough, this pernicious manipulation of racial and ethnic di-visions does not daunt Petras' optimism. He simply dismisses the significance of these factors by predicting an imminent victory of socialist consciousness over divisive, racialist forms of thought. In fact, Petras believes that state capitalism has actually hastened the arrival of the day when a revolutionary alliance will seize state power by heightening the contradictions inherent in the capitalist mode of production. We fear that Petras has let his revolutionary fervor substitute for an analysis of the objective trends and has fallen victim to the common marxist fallacy of underestimating the power and persistence of racial and ethnic variables in the process of social development.

While John Saul would agree with Petras that racial and ethnic conflicts are most often manipulated by in-ternally and externally dominant classes, he does not believe that it is heuristically valuable to dismiss them as epiphenomena. In an article entitled "The Dialectic of Class and Tribe,"[49] Saul observes that "Marxist and other progressive writers on Africa generally approach the issue of 'tribalism' as one would approach a mine-field . . . ."[50] The hesitancy is understandable but Saul rhetorically asks ". . . can anyone doubt that Marxists have barely begun the kind of analysis of ethnicity which is required?"[51] With a detailed know ledge of events in Eastern and Southern Africa in the postcolonial era, Saul fully appreciates that tribalism is a very real factor shaping political processes on the continent. Other marxists such as Archie Mafeje and Richard Sklar, have

seen this too but have resorted to "class-reductionist
approaches." They argue that the politicization of
ethnicity is a simple manifestation of the competitive
maneuvering of local, dominant classes. This is a useful
insight but the explanation is too pat. As Saul points
out, it does not deal with why these ethnic variables are
so readily available for manipulation by the local political elite nor does it explain the obvious relationship
between imperialism and politicized ethnicity.

To understand the availability of ethnic variables
Saul turns to the nature of capitalism's uneven development in peripheral societies. In countries such as Kenya,
Uganda, Tanzania, and Mozambique the economy is based on
what Saul refers to as the "articulation of modes of
production." Pre-capitalist and capitalist modes of
production exist side-by-side. Indeed, the underdeveloped
character of capitalist expansion reinforces the persistence of traditional subsistence economy. The impoverished
tribal communities depend upon the wages sent home by
workers in the cities and the semi-proletarian workers
depend upon the services provided by their kinship group
in times of unemployment, sickness and old age. South
Africa has taken the relationship to a monstrous extreme
and given formal recognition to the system of super-
exploitation in its policy of separate development. A
worker's participation in the modern industrial economy
and urban community is fixed by law to the period of his
or her profitable employment. As soon as this has ended,
he or she is shipped back to the "homeland." Needless to
say, tribal identification remains strong. This does not
mean that its content is primarily traditional or politically regressive.

New institutions and symbols of ethnicity are con-
tinuously being created in response to new challenges.
Sometimes these can be quite progressive and, in special
circumstances, downright revolutionary. Ethnicity,
because of its relationship to the productive structure,
can be used to express (1) class conflicts and/or (2)
center-periphery contradictions. In short, marxists
should not, Saul advises, turn their heads in disgust
everytime they hear the tribal drums beating. Saul
appreciates the difficulty of the problem, but marxists
must explore the link between tribalism and demands for
social justice even if the probability is high that
politicized ethnicity will lead to a blind, often bloody
alley.

It is currently popular to suggest that the limitations of dependency theory and theories of non-capitalist
development can be overcome by a return to orthodox

marxist theory. This we fear is a falsely comforting
notion. Ironically, traditional marxism has not stimulated research into the actual development of the working
class or the conditions under which the working class
adopts socialsim as its ideology. In fact, the very idea
of social determinism in marxist science refer to the
inevitable development of the proletariat.[52] Moreover,
orthodox marxism has simply assumed (1)that the working
class constitutes the marjority in capitalist societies
and (2) that it has a natural interest in socialism.
Thus, what is viewed as problematical within the traditional framework is the process whereby the proletariat is
transformed into a political class--the transition of a
class-in-itself to a class-for-itself. Consequently, the
key questions concern the type of political leadership
required for this task, in brief, the role of the party.
The vanguard role of the working class and its party
within the socialist movement largely goes unquestioned.
    Revisionist marxists have challenged both of these
basic tenets of orthodox marxist theory. The work of
Adam Przeworski[53] and Stanley Aronowitz[54] is most
noteworthy in this regard. According to them, the class
structure of a capitalist society is not a given. They
reject the idea that the proletariat is necessarily the
majority and the revolutionary class simply awaiting
spontaneous organization or activation by the party. No
such pat solution to the problem of socialist revolution
is possible. Classes qua historical actors emerge as the
product of an indeterminate process. Which classes or
groups will emerge as the key actors in the struggle for
social change can not, they argue, be known beforehand.
Moreover, according to Przeworski, this indeterminate
process of class formation is not linear. Classes once
organized do not necessarily remain that way. Classes are
perpetually organized, disorganized and reorganized in the
course of political struggles in which "multiple historical actors attempt to organize the same people as class
members, as members of collectivities defined in other
terms, sometimes simply as members of 'the society.'"[55]
Furthermore, these political struggles are conditioned not
only by objective economic relations but by the totality
of economic, political and ideological relations in the
society. The implications of these revisions of orthodox
marxist theory are far-reaching as indicated by Aronowitz.
"The notion that ideological and political relations as
well as the struggle of immediate producers (whether
mental or manual workers) are possible bases for class
formation opens a new terrain for the theory of historical
change. Further, the idea that more than one outcome may

result from a given set of struggle undermines the concept of historical inevitability with which Marxism has been obsessed for more than a century."[56]

Przeworski's conclusions in regard to some of the unintended and paradoxical consequences of attempts to build a socialist movement are especially relevant to Third World societies which generally lack strong working class institutions and a tradition of popular partici- pation in politics. Socialist forces having made the decision to participate in bourgeois political institu- tions, i.e., electoral politics, must organize a party capable of mobilizing the maximum electoral support. The resulting preoccupation with winning votes will, he argues, led the socialist party to de-emphasize class issues which divide the electorate and emphasize issues around which a broad constituency can be mobilized. The immediate goal of the party thus becomes the projection of an image of itself as a national institution representing the interests of all oppressed people, not just the working class. This of necessity involves a de-emphasis of class conciousness and re-newed attention to other bases of collective identification, i.e., race, sex or religion. The general effect, Przeworski concludes, is one of reinforcing a classless image of society.

Przeworski's analysis is based on the historical experience of social democratic parties in Western Europe. The forces which undermine class-based movements in under- developed countries are more numerous and difficult to handle. In most Third World countries, the working class, the supposed vanguard of a socialist movement, is both a small and privileged section of the total population. The petty bourgeoisie which leads the fierce competition for control of the postcolonial state is also weak and, more importantly, given to fractionalization. Very often members of the petty bourgeoisie will adopt socialist ideology as a means to mobilize the nation in opposition to foreign exploitation. However, as John Saul's theory of Third World politics predicts, internal power struggles will most likely lead petty bourgeois politicians to de-emphasize class and reassert the primacy of ethnic and/or racial identification. Although Saul argues that the activation of tribal sentiment can be " . . . inte- grated into a radical project in such a way as to permit the release of their energies and their anti-imperialist content into a broader terrain of struggle, a terrain where it can overlap positively with both class and national interpellations"[57], more often it represents "the lowest common denominator of petty-bourgeois aspir- ations".[58]

In the following chapters we will utilize these theoretical insights to examine the transformation of the development strategy of the postcolonial state in Guyana. In particular, we will analyze the process of radical-ization which began in the late sixties and culminated in the nationalization of the property of Booker McConnell, Ltd., the oldest and largest expatriate company in the country. The ruling party, the PNC, has declared that the recent and dramatic policy shifts are due to the socialist revolution it has chosen to lead. Critics of the PNC regime have rejected the Goverment's socialist pronounce-ments as distinctly self-serving and suggested that a corrupt, petty bourgeois dictatorship is emerging in Guyana. Clive Thomas makes this point quite forcefully while criticizing the theory of non-capitalist develop-ment.

> This study shows that the period of so-called non-capitalist development can mask—and fre-quently has masked—anti-national and anti-democratic developments. Unless, therefore, we keep as our principal focus the stages of the class struggle, unless we pay close attention to the scope and extent of development of working class institutions and the independent roads available for the working class to come to power, we cannot deal with the problem posed at the beginning, i.e., can socialism be built at this stage of Guyana's development?[54]

The question is excellent and all too often overlooked in discussions of socialism in the Third World. We will attempt to answer it in our analysis.

In chapter two we examine the evolution of Guyana's racialized class structure and the early attempts to organize a united working class movement. This historical background is essential to our analysis because Guyana's first mass-based independence movement adopted a socialist orientation and laid its foundation in the country's trade union movement. We also provide a detailed analysis of the post-war modernization of Bookers, the major sugar producer in the colony, since the struggle for indepen-dence was in large measure fought out on the estates.

Chapter three focuses on the imperialist response to the threat posed by the socialist oriented· independence movement uniting both Blacks and Indians. Bookers, a British expatriate company, sought to make itself invisi-ble through a process called Guyanization—wherever possible, white expatriates were replaced with local personnel. The British Colonial office exercised the

state's coercive powers and dispatched troops to halt the
march towards independence. The Americans, hysterical
over the spread of "communism" in the Western Hemisphere,
chose to subvert the socialist movement by destroying its
base in the trade union movement. The tenuous unity of
the independence movement could not withstand the external
pressure which exasperated internal factionalism and split
along racial lines. Burnham, a temporizing socialist,
proceeded to mobilize a Black base of electoral support.
Jagan, a devoted marxist-leninist, followed suit and built
his base of support in the Indian population, particularly
among the Indian sugar workers.

In chapter four we analyze the racialization of the
development strategy pursued by the postcolonial state
under the leadership of Forbes Burnham. In order to
capture control of the state, Burnham's party, the PNC,
had to form an alliance with foreign and local capitalists. A few socialist phrases remained but they were used
to garnish the usual capitalist development plan for Third
World economies. When the magic of the market failed to
deliver the goods, the PNC began its leftward odyssey
which climaxed in its adoption of marxism-leninism.
Particular attention is given to the content of the
Government's new socialist partnership with labor.

Chapter five is devoted to a detailed analysis of the
PNC's policy in the sugar industry. This, as we pointed
out in the beginning, is of vital significance since
developments in the sugar industry are so closely linked
to the overall pattern of economic, political and racial
dynamics in the country. If relationships in the sugar
belt are characterized by exploitation, repression and the
manipulation of racial hostility, then life in Georgetown,
New Amsterdam and Linden are similarly degraded. Socialism cannot be built if the working class remains divided
along racial lines and the two major leftist parties stay
at each other's throats. Therefore, we need to find out
if the Government's decision to nationalize the sugar
industry (the stronghold of East Indian resistance)
represented a renewal of the long severed partnership
between the PNC and the PPP in pursuit of a socialist
revolution. In order to make this determination, we
examine the Government's sugar policy between 1964 and
1977. We deliberately extend our analysis beyond the date
of nationalization so that we might see how state ownership and control have affected the structure of the
industry, particularly, the relationship between the PPP's
powerful union in the industry and the Government. In
sum, is the sugar industry now a site of socialist reconstruction or does it remain a fiery cauldron of racial,

political, and class conflicts?   Chapter six draws the various strands of our analysis together and offers a necessarily tentative conclusion regarding Guyana's future development.

NOTES

1. J.S. Furnivall, Colonial Policy and Practice: A Comparative Study of Burma and Netherlands India (London: Cambridge University Press, 1948), P. 304. Emphasis added.

2. M.G. Smith, The Plural Society in the British West Indies (Berkeley, California: University of California Press, 1965). Leo Depres, "Differential Adaptations and Micro-Cultural Evolution in Guyana," Southwestern Journal of Anthropology 25 (1969): 19-43; Cultural Pluralism and Nationalist Politics in British Guiana (Chicago: Rand-McNally, 1967); Harry Hoetink, The Two Variants in Caribbean Relations: A Contribution to the Sociology of Segmented Societies (London: Oxford University Press, 1967); Neville Layne "The Plural Society in Guyana" (Ph.D. dissertation, University of California, 1970).

3. Kathrine West, "Stratification and Ethnicity in 'Plural' New States," Race 33, 4 (1972): 493.

4. R. T. Smith, "Race and Political Conflict in Guyana," Race 12, 4 (1971): 417.

5. Ibid., p. 427.

6. Andre Gunnar Frank, Capitalism and Underdevelopment in Latin America (New York: Monthly Review, 1969), pp. xi and xv.

7. Frank, Capitalism and Underdevelopment; Theotonio dos Santos, "The Structure of Dependence," American Economic Review, 60 (May 1970): 231-236; F. H. Cardoso, "The Consumption of Dependency Theory in the United States," Latin American Research Review, 12, 3: 7-24; Paul Sweezy, "Modern Capitalism," Monthly Review 23 (June 1971): 1-10.

8. Frank, p. 10.

9. Ibid., p. 9.

10. S. Amin, Unequal Development (London: Harvester, 1974); A. Emmanuel, Unequal Exchange (New York: Monthly Review, 1972); Walter Rodney, How Europe Underdeveloped Africa (Washington, D.C.: Howard University Press, 1974); Norman Girvan, Corporate Imperialism: Conflict and Expropriation (New York: Monthly Review, 1976).

11. Frank, p. vx.

12. Quoted in A. G. Frank, Lumpenbourgeoisie and Lumpendevelopment: Dependency, Class and Politics in Latin America (New York: Monthly Review, 1972), p. 1.

13. Quoted in Frank, ibid., p. 7.

14. Andrew Zimbalist, "Synthesis of Dependency & Class Analysis," Monthly Review 32 (May 1980): 27-31.
15. In James Petras, Critical Perspectives on Imperialism and Social Class in the Third World (New York: Monthly Review, 1978), pp. 13-62.
16. Ibid., pp. 33-34.
17. Ibid., p. 39.
18. Ibid., pp. 36-37.
19. Ibid.
20. Ibid., p. 48.
21. Ibid., p. 44.
22. Ibid., p. 45.
23. Ibid., p. 47.
24. Ibid., p. 46.
25. Ibid., p. 48.
26. Ibid., p. 50.
27. Ibid.
28. V. Solodovnikov and V. Bogolovsky, Non-Capitalist Development: An Historical Outline (Moscow: Praeger Publishers, 1975); Jay Mandle, "Problems of the Non-capitalist Path of Development in Guyana and Jamaica," Politics and Society 7, 2 (1977): 189-197; Ralph Gonsalves, "Towards a Non-Capitalist Development Strategy," Paper presented at the Queens college, CUNY, conference on Underdevelopment and Development in the Black World, 8 May 1980.
29. John Ohiorhenuan, "Dependence and Non-Capitalist Development in the Caribbean: Historical Necessity and Degrees of Freedom," Science and Society 18 (Winter 1979-1980): 389.
30. Jay Mandle, p. 196.
31. Charles Bettelheim, "Dictatorship of the Proletariat, Social Classes and Proletarian Ideology," Monthly Review 23 (November 1971): 62.
32. John Ohiorhenuan, p. 408.
33. Quoted in John Saul, The State and Revolution in Eastern Africa (New York: Monthly Review, 1979), p. 3.
34. Harry, Magdoff, "Is there a Non-Capitalist Road?," Monthly Review 30 (December 1978): 1.
35. Ibid., p. 3.
36. Ian Roxborough, Theories of Underdevelopment (London: Macmillian Press, Ltd., 1979), pp. 146-147.
37. Ibid., p. 142.
38. Ibid., p. 138.
39. Ibid., p. 133.
40. Ibid., p. 135-36.

41. James   Petras,   Critical Perspectives in Imperialism and Social Class in the Third World   (New   York: Monthly Review, 1978), pp. 84-102.
42. Ibid., p. 85.
43. Ibid., p. 87, emphasis added.
44. Ibid., p. 86.
45. Ibid., p. 87.
46. Ibid., p. 99.
47. Ibid., p. 100.
48. Ibid., p. 101.
49. John Saul, "The Dialectic of Class and Tribe."   Race and Class 20 (Spring 1979): 347-372.
50. Ibid., p. 347.
51. Ibid., p. 371.
52. D.C.   Cox,   Caste, Class and Race, A Study in Social Dynamics (New York: Modern Reader Paperbacks, 1970).
53. Adam Przeworski, "Proletariat into a Class: The Process of Class Formation from Karl Kausty's The Class Struggle to Recent Controversies,"   Politics and Society, 7 no. 4 (1977).
54. Stanley Aronowitz, The Crisis in Historical Materialism: Class Politics and Culture in Marxist Theory. (New York: Praeger Publishers, 1982).
55. Przeworski, p. 373.
56. Aronowitz, p. 77, (original emphasis).
57. Saul, p. 368.
58. Saul, The State in East Africa, p. 9.
59. Clive Thomas, "Broad and Justice: The Struggle for Socialism in Guyana,"   Monthly Review 28 (September 1976): 28.

Obstacles to the formation of a working class movement in Guyana

The formation of the Guyanese working class is a terribly neglected aspect of the country's history. There are to our knowledge only two books which have been devoted to the vital questions of the formation of the working class in Guyana: Aston Chase's The History of Trade Unionism in Guyana, 1900 to 1961 and Walter Rodney's A History of the Guianese Working People, 1881-1905. The latter was intended to be part of a two volume study but it will not be completed since Rodney was killed before the publication of the first volume. Harold Lutchman has written an insightful article on current trends in the Guyanese trade union movement[1] but beyond these works is very little.

This oversight of working class history is particularly unfortunate since a radical socialist party, the Peoples' Progressive Party, organized Guyana's independence movement in the late 1940s and has remained a major factor in its political development ever since. There are of course different interpretations of the correct role of a party within a socialist movement but most marxists would agree that it represents the political vanguard of working class interests. The PPP has always claimed to be the vanguard of the Guyanese working class. The Peoples' National Congress, the ruling party has more recently claimed this vanguard role following its adoptation of marxism-leninism as its official ideology. This predominance of marxist ideology in Guyanese politics and the persistence of deep racial divisions makes it even more unfortunate that so little attention has been focused on either the objective or subjective development of the working class in the country.

Sugar as King

From the 16th century well into the 20th century, Sugar was King in Guyana. Everything was subordinated to the interests of first, a resident class of sugar planters and later, expatriate owners of London-based, joint stock companies. To ensure its monopoly of capital, land and labor, the planter class excluded all other classes from meaningful political participation. Effective state power was securely kept in the hands of the Governor and the Colonial Secretary appointed by the Crown and the Combined Court controlled by planters or their representatives.

The economic power of planter class was awesome. Almost
all available work was directly or indirectly tied to
sugar. The political power of the planter class guaranteed that the colony would remain little more than a
string of sugar plantations.

Africans were brought to Guyana to work as slaves
on the plantations in the 1600s after the Europeans discovered that the indigenous, Amerindian population could
not satisfy their need for a large labor force. Aside
from early trading contacts with the Europeans and their
later employment as trackers of run-away slaves, the
Amerindians remained outside of the colonial political
economy which developed along Guyana's coastal strip.
Until the abolition of slavery in 1833 Africans supplied
all of the unskilled and much of the skilled labor required for sugar production. European planters hired other
whites to perform management and supervisory functions and
certain categories of skilled labor. An unmistakable
color-caste hierarchy was firmly in place before emancipation. Wealth, power and privilege were co-terminus
with white skin. While whites held the power of life and
death over Black workers no Black man stood on equal
footing with any white. Nevertheless, the Europeans
justified their superordination in terms of their wealth,
superior culture and education, not in terms of race.
Still no one could miss the perfect correlation between
race and status. Colored or mixed persons, initially the
product of illicit sexual relations between white men and
defenseless Black females, grew in number and formed a
convenient buffer between the planter class and the
oppressed. Frequently the beneficiaries of an education
abroad, jobs as overseers or even material inheritance,
the Coloreds closed ranks against their less fortunate
Black brothers and sisters and vainly sought full acceptance in the status group of their white fathers. The
abolition of slavery added greater complexity to the
color-caste hierarchy through the institution of
indentureship and a degree of fluidity through the creation of a class of free wage earners.

Indentured Immigration

To a greater extent than other Caribbean producers,
Guyanese planters saw the abolition of slavery as a deadly
assault upon their class interests. Because Guyana's rich
coastal lands had to be reclaimed and constantly defended
against the sea, its cost of sugar production was much
higher than that of its neighbors. In addition, Guyana

had a vast interior, rich in natural resources such as timber, gold, and diamonds. In order to survive, Guyanese planters had to ensure a large supply of cheap labor submissive to authoritarian discipline. To accomplish this the planters had (1) to subvert the operation a free labor market through which workers could seek to improve their wages and condition of employment and (2) to prevent the development of alternative economic sectors in the interior. The ways in which the planters sought to accomplish these goals and the consequences for the process of class formation in Guyana is the subject of Walter Rodney's last major work. In the following discussion we will rely heavily on his historical analysis although, as we will later make clear, we reject the theoretical conclusions he reaches.

Before the end of apprenticeship released the ex-slaves from enforced labor on a given estate, shiploads of indentured workers arrived in the colony. State-aided indentureship remained a vital weapon in the planters' struggle with free labor until the barbaric practice was finally ended in 1917. Free African labor exercising its rights to seek better employment and to withhold its labor power immediately threatened the profitability of the Guyanese sugar industry. In 1842 the new wage earners demonstrated their determination to use this vulnerability of the planters to defend their standard of living and to seek better wages. The planters temporarily retreated from their organized effort to drive down wage rates but they intensified their plans to increase indentured immigration. When free African labor mounted another major strike in 1848 the planters were ready. Indentured workers replaced free labor and the strike was a failure. To add insult to injury, the ex-slaves, as taxpayers were helping to finance the importation of bound labor which would continue to depress their wages and reduce job opportunities.[2]

From the outset, indentureship was denounced as a new form of slavery. An indentured immigrant arriving in the colony was bound to work on a particular estate, at a fixed wage, for a predetermined number of years. He or she would later be entitled to free repatriation. Having entered this contract, the immigrant surrendered all personal freedom to a planter who at best offered paternalistic care. For six days a week, 15 to 20 hours a day, over a period of 5 to 10 years, indentured workers toiled without the option of refusing an assignment, winning a pay increase or seeking a better job. Failure to report to work was considered "desertion" and legally prosecuted as a criminal offense. Every year, thousands of

indentured workers were dragged before the local magis-
trates who resolutely defended the planters' interests by
imposing extravagant fines and excessive prison sentences
for the slightest breach of contract. Free Africans
denounced the semi-feudal institution which subverted the
operation of free labor market and correctly pointed out
that indentured immigrants labored under the same conditions they had experienced in chattel slavery.

Planters variously claimed that indentureship was an
indispensible source of cheap labor or that it was more
expensive than available free labor.   Where the truth
actually lies need not concern us here. What is important
is that the planter class successfully used indentured
immigration to break the back of a new class of free wage
earners. Wage rates for bound and free labor which were
set before the end of apprenticeship in 1838 remained in
force until indentureship was abolished in 1917.[3] By
the mid-1870's, three district categories of estate labor
had developed: (1) indentured immigrants under contract,
(2) time-expired immigrants still living and working on
the estates, and (3) free labor residing in nearby villages working part or full time on the estates. The
important point to remember, according to Rodney's analysis, is that the fate of all categories of labor was
ultimately determined by the plight of contract or bound
labor.   Although unemployment and underemployment were
grave problems and mechanization steadily reduced the
labor needs of the industry in late 19th century, sugar
planters continued the importation of indentured workers
and the colonial state subsidized one-third of the cost.

Color-Class Hierarchy

In addition to these obstacles to the development of
the working class, indentureship also reinforced the
colony's social hierarchy based on race, class and culture. To the existing pecking order of Europeans/Whites,
Mixed/Browns, and Africans/Blacks, the newcomers, Portugese, Chinese and East Indians, were added.   Over
fifty thousand Africans and West Indians from other
territories were also brought in but they were quickly
absorbed into the local Black population.

Portugese. The first shipment of Portugese arrived
in 1835. By 1880 over 60,000 had been brought in as
contract laborers. Their experience is especially useful
for pointing out some of the peculiarities of the system
of ethnic and class stratification in the colony. Portugese are European and white but they have never been

regarded as such in Guyana because they entered as indentured workers. The indentured Portugese may have looked like the plantation owner and overseer but the free African enjoyed greater rights and mobility. This aberration in the color-class hierarchy did not last long however. Planters were not satisfied with the Portugese as a replacement for ex-slaves because their supply was limited and they could not sustain the rigors of heavy agricultural labor in a tropical climate. Only a trickle entered the colony after 1860.

More importantly, those already present were in mass transformed into a commercial middle class. Planters cancelled labor contracts before their date of expiration and made available credits which enabled the Portugese to gain a monopoly on retail trade in the countryside and cities. With assistance from the colonial state in the form of land grants, credits and licenses, the Portugese moved into provision farming, woodcutting, gold mining and other minor industries. By the late 19th century, the Portugese had consolidated themselves as a distinct ethnic group and the dominant segment of the commercial middle class. As such they came into increasingly sharp conflict with the planters who saw economic diversification which the middle class championed as a challenge to their monopoly of land, labor and capital.

Chinese. The importation of indentured Chinese laborers for the sugar estates was also a failure. The first group of Chinese immigrants entered Guyana in 1854. Few would follow after 1866. In 1860 the number of indentured Chinese workers reached its peak of 10,000; by 1879 it had dropped to 6,000. The decline in the Chinese percentage of the population was, according to Olando Paterson, irreversible.[4] There were only 2,000 Chinese in the colony in 1911. The small number of Chinese and the gross disproportion of men to women help to explain their unique pattern of accomodation. Unlike the Portugese and Indians, the Chinese never tried to maintain their ethnic identity and frequently intermarried with other racial groups. In terms of occupation, the Chinese also managed considerable flexibility. After their period of indentureship they strictly avoided agricultural work and sought their livelihood in the urban areas. Many went into retail trade but without the incentives given to their rivals in the business they "...had to be satisfied with the pickings left by the Portugese traders."[5] Others branched out into various skilled urban occupations. A few made it into the civil service and professions.

　　　East Indian and Black Labor. Unsatisfied with their efforts to substitute Portugese and Chinese labor for free wage labor, plantation owners turned to India as the source for indentured immigration. Throughout severe and protracted crises in the international sugar market, mechanization and amalgamation in the Guyanese industry and a desperate crisis of unemployment facing the creole labor force, sugar producers continued to import indentured workers from India. Because the vast majority chose to remain, Indians replaced Blacks as the single largest ethnic group in the colony. More importantly, the steady increase in the Indian population disarmed free Black labor in its conflict with the planter class.

　　　Initially, Indians replaced Black workers in the performance of the least desirable and poorest paying jobs. Blacks took advantage of their experience and skill and concentrated on the better paying and less back-backing work in the factories and fields. Consequently, although Blacks and Indians worked on the same estates, the racial allocation of various categories of labor prevented their development into an integrated work force. Black workers highly praised for their skill with the cutlass and shovel, worked in all "Creole gangs" as cane-cutters and shovel men. Meanwhile Indians worked in "coolie gangs" weeding and manuring the fields. Despite this advantage, Blacks who remained in the industry still could not earn a decent living as sugar workers. The twin pressures generated by the sharp decline in sugar prices starting in 1884 and the strike breaking capacity of bound, Indian labor often drove the wages of free workers below that fixed by law for indentured workers. With the exception of a few highly skilled factory jobs, Blacks had to abandon the industry and seek their livelihood elsewhere. Indians thus became the overwhelming majority of workers in the sugar industry--the most important sector of the colonial economy.

　　　Forced out of the sugar industry, Blacks had few economic alternatives. Immediately after slavery, they had struggled to transform themselves into an independent peasantry but the planters had blocked them at every turn. At their behest, the Governor refused to release crown lands along the coast or in the interior for economic development. None of the public treasury went towards the creation of an infrastructure to support peasant cultivation. Meanwhile, the planter class refused to sell, lease or rent any of its vast land holdings except under the most extreme terms of exploitation. As a result, Blacks had to exhaust their limited savings

to purchase lands they subsequently could not afford to
properly maintain. Many plots were abandoned while others
were auctioned off by the state to cover unpaid taxes and
fees. The free African villages which survived were the
sites of economic stagnation. Embittered by their
defeat both as wage workers in sugar and as peasant producers, Blacks increasingly turned away from agricultural
labor.

As yet another crisis gripped the sugar industry in
the 1880's, many Black males headed for the hinterland and
employment in the newly expanding industries there. Some
hoped to strike it rich as gold or diamond prospectors.
Others worked as woodcutters or balata bleeders. Whatever
the particular pursuit, few could piece together a decent
living. The expansion of the gold, diamond and timber
industries in the 1880s and 1890s marked the emergence of
Guyana's modern political-economy. However, according to
Rodney, this development was not enough to transform the
African labor force into a modern working class. Most
African workers were locked into a semi-proletarian position. In Rodney's words villagers "...became instead
potential members of a free labor force and were amenable
to numerous forms of labor."[6] Most male workers migrated between the coast and the interior, working as cane
cutters and balata bleeders, peasant farmers and gold
miners, woodcutters and shovelmen on the estates. Likewise, many of those who went to the urban areas could not
escape the incomplete crystallization of their class
position. Working at the most menial jobs, they too
migrated between the city, countryside and hinterland
picking up various, casual jobs.

When the bauxite industry developed during World
War I, Black men flocked to the area located sixty miles
up the Demerara River where deposits had been discovered.
The physical and social isolation of the mining communities was compensated for by the fact that some Blacks
could find good paying jobs and steady employment as
bauxite miners. Life in the mines and the towns was run
in accordance with the companies' strict apartheid policy.
Only Blacks were hired as miners and all the top managerial, supervisory and technical positions were reserved for
white expatriates. Indians were not employed as miners
thus adding a new dimension to the colony's ascriptive
system for the allocation of work and social rank.
Bauxite miners became the local aristocracy of labor but
relatively few Blacks could find employment in the
industry. Moreover, the foreign owners were not interested in developing the forward and backward linkages to the
rest of the colonial economy necessary for genuine

economic development. The bauxite industry remained an enclave in the economy.

The crisis in the sugar industry during the 1880s and 1890s also had tremendous impact upon the process of class formation in the Indian population. With wages for free labor declining and unemployment rising, many time-expired indentured immigrants decided to exercise their right to free repatriation to India. Needless to say, this practice was not conducive to the long-term interests of the planter class. Despite the slump in the sugar market, the estates had to ensure that there would be a large and cheap labor force on hand for the better days ahead. To meet this challenge and to escape the expense of mass repatriation, the planters reluctantly decided to promote Indian peasant farming. This time the colonial state was enlisted to assist in the formation of an independent peasantry.

In lieu of return passage, the colonial state offered time-expired immigrants free grants of land. Unfortunately, this land was frequently taken over from Africans who were currently renting or leasing the land.[7] In 1897 the Royal West Indies Commission recommended that the potentially vast resources of Crown lands finally be opened up for agricultural and commercial development. Africans did not for the most part benefit from this change in colonial policy; but, Indian farmers did receive significant grants of land. The planters likewise adopted a policy of renting or leasing estate land not under cultivation to their best Indian workers. The logic underlying the policy was sound. At little or no cost to the planters, small plots of land could be used to reward loyalty and to tie the Indian worker to the plantation almost as tightly as the system of indentureship. A household could supplement the starvation wages paid by the estates with small-scale farming and manage to stay alive.

Most of the Indians concentrated on rice farming which emerged as the third major sector of the colonial economy by the end of World War I. As in the case of bauxite, it developed as the preserve of one racial group. From top to bottom, the rice industry is an Indian operation. Black small farmers have historically concentrated on ground provisions.[8] Not surprisingly, only a few Indians became wealthy as rice farmers, rice millers or landowners. By 1911 less than half of the Indian population lived on the estates but they nonetheless remained overwhelmingly dependent upon wages from employment in the sugar industry. Unable to earn a decent living as either a sugar worker or rice farmer, most

Indian workers had to divide their time and energy between the two. Notwithstanding their greater success at small-scale farming, the class position of most Indian workers remained ambiguous. The typical work trajectory for an Indian worker included a period of indentureship, wage labor in the sugar industry and peasant farming. As in the case of Black workers, the inadequacy of any particular pursuit forced considerable overlap on to most Indian workers.

The development of the rice industry had ramifications beyond the economic realm. Peasant farmers and the emerging middle class provided the material foundation for the revival of "Indian culture" in Guyana. Successful farmers and businessmen provided much of the money and leadership of the movement to preserve elements of Hindu and Moslem culture which served as the basis for the consolidation of the Indian population as an ethnic group. It did not matter that most Indians no longer spoke their native tongue and that most of their rituals and ceremonies had already undone significant creolization under the regimen of the plantation system. The point is that this cultural heritage was used to promote Indian pride and group solidarity. The significant rise in the number of Hindi temples and Moslem mosques during the late 19th and early 20th century is evidence of this process of ethnic group consolidation.

Conversely, Africans eagerly sought to assimilate British culture. Of course the process could never be complete given the crushing poverty of the Black population and the underlying assumption of British colonial policy which held that only a white man could be an English gentleman. The colonial school system was the main agency for socialization into a creolized version of British culture. The fact that the largely state financed school system was controlled by Christian denominations helps to explain why Africans became highly creolized and Indians retained far more of their cultural distinctiveness. Africans saw the schools as the best avenue to social mobility and strongly emphasized educational achievement to their children. Indian workers suspected that the schools were guilty of proselytism and generally kept their children away. The colonial state condoned this practice by exempting Indian parents, as non-Christians, from prosecution under the Compulsory Education Ordinance of 1897. This exemption was not removed until 1933.[9] Consequently, Black people used education as their primary means of entering the middle class. They became school teachers, headmasters, ministers, junior members of the colonial civil service and entered the

legal and medical professions. Meanwhile, the Indian
middle class developed on the basis of traditional petty
bourgeseois pursuits--commercial trade, land ownership and
farming.

Traditional marxist theory does not account for the
pattern of development observed in Guyana. Although the
colony was integrated into the emerging world capitalist
system in the early 17th century and supplied some of the
capital used to finance England's industrial revolution,
Guyana became a peripheral part of the new industrial
order. No large-scale industrial enterprises emerged.
Foreign capitalists ironically fought to maintain "non-
capitalist" relations of production in order to ensure
profits they then invested in modern industries in the
metropolitan centers. Therefore only half of the marxist
prediction of the polarization of a capitalist society
into two antagonist classes holds true for Guyana. On the
one hand, the anarchy of the sugar market and techno-
logical development eliminated inefficient producers and
concentrated capital in the hands of a smaller and more
powerful class of capitalist owners. But, on the other
hand, the supposedly inevitable development of a proletarian majority did not occur. Harold Wolpe explains this
peculiar feature of capitalist expansion in the periphery as follows.

> In certain conditions of imperialist development,
> ideological and political domination tend to be
> expressed not in terms of the relations of class
> exploitation which they must sustain but in
> racial, ethnic, national, etc., terms and, in all
> cases, this is related to the fact that the
> specific mode of exploitation involves the
> conservation, in some form, of the non-capitalist
> modes of production and social organization, the
> existence of which provides the foundation of
> that exploitation. Indeed, it is in part
> the very attempt to conserve and control the non-
> capitalist societies in the face of the tendency
> of capitalist development to disintegrate them
> and thereby to undermine the basis of exploita-
> tion, that accounts for political policies and
> ideologies which centre on cultural, ethnic,
> national, and racial characteristics.[10]

Thus in Guyana, capitalist exploitation did not lead to
the development of a working class majority or the gradual
destruction of racialist forms of consciousness. On the
contrary, "a permanent hybrid of peasant and proletariat"[11] emerged and racial identification remained the
bedrock of collective consciousness.

Within the context of Guyana's colonial political-
economy, the barriers to working class unity were formidable. Workers, shifting back and forth between the
countryside, city and hinterland and working at various,
casual jobs and/or on small plots of land during periods
of slack employment, were not predisposed to develop a
strong sense of "we-ness" and common destiny with other
transient workers. Sugar workers were likewise poor
candidates for proletarian conciousness and organization
despite the large-scale and semi-industrial nature of
sugar production in Guyana. First, the serf-like status
of indentured workers and second, the racial division
within the labor force, virtually precluded the transformation of sugar workers into a power bloc capable of
effectively confronting the planter class.

Needless to say, the importation of Indians as strike
breakers created a reservoir of ill-feelings among Black
workers. Subsequently, the racial system for allocating
work within the industry buttressed the schism by physi-
cally segregating Blacks in the sugar factories and
Indians in the cane fields. Long after leaving the
plantations, Blacks perceived Indians as competitors
"taking the bread out of their mouths." [12] The unfair
competition presented by bound labor fueled the feelings
of injustice Black Guyanese felt living in a society
controlled by white Europeans. Although brought to Guyana
in chains, Black ex-slaves came to perceive the country as
their rightful inheritance. East Indians were, not
without justifications, seen as temporary intruders
manipulated by the planters. When during the late
19th century it became clear that most Indians were not
sojourners but permanent residents claiming their stake in
the country's future, mutual suspicions heightened. The
racial stereotypes used by the colonizers to rationalize
their mistreatment of the labor force were to some extent
accepted by Blacks and Indians. Indians are supposedly
hard-working, niggardly, furtive and clanish. Blacks are
meanwhile cast in the sambo role--easygoing, fun-loving
and improvident. Such racial attitudes run deep in a
society and are amazingly resistant to change. Even today
it is common to hear and Afro-Guyanese complain that "the
coolies" (a common pejorative term for Indians) are taking
over the country, buying up the best property in Georgetown and manipulating the government through the power of
the purse.

It is indisputable that racial and cultural differences have exerted a powerful impact on Guyanese history.
The theoretical debate between those utilizing a cultural
pluralist perspective or marxist framework revolves around
whether or not primary explanatory weight should be

assigned to racial and/or cultural factors or the class
structure. Working within an orthodox marxist framework,
Walter Rodney pointed to the fact that

> Ideological confusion and psychological oppres-
> sion were as crucial to the maintenance of the
> plantation systems as were the administrative
> control and the final sanction of the police
> force. In a heterogeneous society, the impact of
> racist perceptions was obviously magnified, and
> its principal consequence was to hold back the
> maturing of working class unity by offering an
> explanation of exploitation and oppression that
> seemed reasonably consistent with aspects of
> people's life experience.[13]

Partly as a result of this ideological confusion, no
broad-based working class organization emerged in the 19th
century to mobilize workers and challenged the plantocracy. Indeed, within the context of scant economic
opportunities and active manipulation of racial suspicions, Rodney argues that it is remarkable that there
were so few violent conflicts between Blacks and East
Indians. Nevertheless, Blacks and East Indians have only
temporarily been able to see beyond their differences to
unite into a class-based movement for social change.
   Walter Rodney's analysis is on the whole rigorous and
penetrating. However, his insistence upon forcing a
marxist interpretation upon events when it obviously does
not provide the best explanation is disturbing. His
analysis of the 1905 strikes and riots is a good example
of his attempt to square the circle. Faithful to his
careful historical style, Rodney observes that the strikes
of 1905 were leaderless, unorganized and produced no significant gains. However, despite the absence of working
class organization he would have us see in these events
evidence of marxist prophecy. For Rodney, the uprising
demonstrated working class consciousness in advance of
local, objective circumstance probably inspired, he suggests, by contemporary struggles taking place within the
international working class movement. Notwithstanding the
economic backwardness and host of social and cultural
obstacles operative in Guyana, Rodney insists that the
classical marxist model still fits. Why? Because during
the post-emancipation period ex-slaves became "potential
members of a free labor force" and in 1905 those "shot at
Ruimveldt were definitely vanguard workers in a period
when there was no organization for workers."[14]

In our view, terms such as "potential members" and "vanguard workers" simply cloud over the important theoretical and ultimately practical issues involved. Conclusions like the following serve to undermine rather than enhance a conceptual analysis of history: "The social situation in the early 20th century was still fluid. In its analysis, one must utilize the terms 'working class,' 'working people,' and 'the people" to refer (in that order) to entities within larger entities, among which the contradiction were not fundamental."[15] Be that as it may, class conciousness did not leap over the objective and subjective barriers in 1905. Those involved in the strikes and riots were primarily Black. Even the sugar workers who joined the demonstrations were Africans-- Indian sugar workers stayed away. The spontaneous uprisings dissolved into confusion with workers accusing each other in order to protect their own skins. Rodney's analysis unduly rests upon what might have been--a vision colored by doctrinal faith. Aston Chase's conclusions in regard to the failure of the 1905 uprisings are, we feel, more realistic: "The principal material basis for the development of proletarian class conciousness such as large scale industry were not present."[16] Objective conditions do not by a long shot guarantee the development of revolutionary consciousness but it is a necessary condition.

Working Class Struggles Between 1905 and 1939

Walter Rodney's study of the formation of the Guyanese working class covers the years 1881-1905. After this time we must rely upon Aston Chase's analysis of the trade union movement which covers Guyanese labor history up to 1961. Like Rodney, Chase was actively involved in Guyanese politics and analyzed the organizational development of the working class from an orthodox marxist perspective. From Chase we learn that socialist ideology made an early appearance in the Guyanese trade union movement (TUM).
The very first trade union in the colony adopted a socialist program. The British Guyana Labor Union (BGLU) was founded in 1919 by Hubert Critchlow in the wake of the widespread strikes of 1917. War induced inflation and the persistence of starvation wages had driven workers into open confrontation with employers and the colonial state. This time Indian workers joined the disturbances and sought out the leadership of Critchlow, a dynamic, Black labor activist. During its first year, 7,000 workers representing an impressive cross-section of labor joined up and paid membership dues. An official BGLU publication

dated 1922 explained that the goal of the organization was
the creation of a collectivist state and proletarian
dictatorship. In this respect, strikes and collective
bargaining would take a back seat to direct political
mobilization to win such victories as the nationalization
of railways, land and the means of production and exchange. According to Chase the document reflected a
sophisticated grasp of socialist doctrine and and was
probably written in England. Chase concludes that "The
events will show that the union did not uncompromisingly
pursue the class struggle and that its early founders were
not imbued with a scientific analysis and understanding of
the class struggle."[17] At the time of this writing in
1961, Chase held a marxist-leninist position and therefore
believed that the failure of Guyana's past socialist
ventures was due to an unscientific approach to the class
struggle. The more obvious and convincing explanation is
that local conditions did not support such an ambitious
goal. In the early 1920s, Guyanese workers were deeply
divided racially, disfranchised of all political rights
and without a tradition of working class organization. In
addition, the colonial state was unambiguously the repressive instrument of the planter class. In this environment it is not surprising that the first, largely
Black trade union quickly lost its revolutionary fervor.
Within a very few years, BGLU membership declined sharply
and the union was transformed into a springboard for the
personal and political ambitions of its middle class
leadership.

Throughout most of Guyana's history, sugar workers
have represented the single largest labor force in the
country. While wage labor outside the sugar industry was
scattered in small-scale businesses, workers in the sugar
industry experienced similar conditions of labor and
exploitation on large-scale estates employing several
hundred workers. Nonetheless, organizing sugar workers
was an extremely difficult task. In fact very little by
way of trade unions could be achieved before the system of
indentureship was eliminated in 1917.[18] Isolated on the
estates and disfranchised of all rights, the Indian sugar
workers relied upon the Immigration Agent-General and
prominent Indian merchants and lawyers to intercede on
their behalf. In light of this history of indentureship
and the ingrained pattern of racial competition, Chase
correctly points out that "considerable credit" is due to
Hubert Critchlow and the BGLU for attracting the support
of Indian workers.[19] Nevertheless, the union primarily
represented the interest of Black workers in Georgetown.[20]

The first major and successful attempt to organize sugar workers occurred in 1937 under the charismatic leadership of Ayude Edun. Edun was inspired by an eccentric and utopian vision in which Britain would become the center of a new, rational world order wherein people would be allotted to fixed social strata on the basis of "accurate scientific statistics" As in Plato's Republic the intelligentsia would rule while man-power citizens would perform the labor nature had best equipped them for. Edun's vision was basically conservative--he organized the Man-Power Citizens Association (MPCA) in the Guyanese sugar industry as "an alternative to Communism, Facism, Ghandism, and sundry other 'isms'."[21] The union enjoyed a brilliant start. Sugar workers saw the union which fought for better working and living conditions on the estates as their long awaited savior. A strike in 1939, in which four workers were murdered by the authorities, resulted in the recognition of the MPCA as the first trade union in the sugar industry. Between 1938 and 1943 over 20,000 sugar workers joined the organization. It was beyond question the largest and most powerful trade union in the colony. The hopes and trust of sugar workers were however quickly betrayed. Immediately after extending formal recognition to the MPCA, the Sugar Producers Association (SPA) proceeded to bribe its foremost leaders. Henceforth, the MPCA functioned as a pliant and dependable company union.[22]

Thus by the start of World War II the Guyanese working people had scored no lasting victories. When the Moyne Commission came to Guyana in 1939 to investigate the causes of the intense labor rioting which had swept the entire British Caribbean during the Great Depression, it found working and living conditions only slightly removed from those of slavery. Wages were abysmal, unemployment was rampant, and workers' attempts to organize were still being mercilessly opposed by employers with the dependable assistance of the state. Meanwhile the colonial state refused to finance any system of unemployment relief or public dole. In Guyana, the legal framework for labor relations in 1939 remained the Employers and Servants Ordinance of 1853. There was not a single collective bargaining agreement in the country.

Militant protest and reformist policy

Despite the many defeats of early attempts to organize Black and Indian workers and the failure of working class consciousness to surmount the racial divide, marxist

theory emerged as the principle ideology of the independence movement in the post World War II period. The two
major nationalist leaders, drawn from the Black and Indian
populations respectively, agreed that independence was
primarily a means to achieve their ultimate goal--the
creation of a socialist society. Inspired by the resur-
gence of militant protest in the colony, the nationalist
leaders decided to concentrate their political mobilization within the trade union movement. They would
simultaneously confront capitalist exploitation and
colonial oppressions. Their immediate aim was the development of a political-trade union complex powerful enough
to wrest state power from the colonial establishment.

The Great Depression, World War II and the rapid
decline of Britian's economic and military might created a
series of major dislocations which fired mass defiance in
her colonial territories. Spontaneous outbursts of
violence, militant trade unionism and nationalist movements all indicated on imminent threat of radical change
if reformist policy was not immediately adopted. The
Colonial Office, often in opposition to powerful colonial
interests, astutely set out to direct the enormous
democratic impulses unleashed by the Depression and World
War. Accepting the inevitability of decolonization,
imperial policy was henceforth designed to foster the
development of local institutions and an indigenous class
which would ensure a favorable investment climate when
independence was finally granted. Programs of educational
reform, slum clearance, public health and housing were
either announced or immediately commenced. Plans and
programs to diversify agriculture and industry were given
public prominence.[23]

One of the most important recommendations of the 1939
Moyne Report published after WWII pertained to the run-
down and outmoded Caribbean sugar industry. Both the
quota and price paid for Caribbean sugar were to be
significantly increased in order to make possible the
modernization of the industry and the improvement of the
deplorable living conditions of sugar workers. Indeed,
this large U.K. subsidy was made contingent upon the
creation of a labor welfare fund for sugar workers.[24]
The Moyne Report also urged drastic reform of the colonies' diaconian system of labor relations.

In the opinion of the Commission's members, it was
time for the colonial state to play a more neutral and
progressive role in the struggle between labor and capital. Instead of unabashedly serving as the ultimate
coercive instrument of colonial employers, the Report
recommended that the state establish official boards to

fix wages, institute unemployment insurance schemes in the
larger colonies, supervise health standards and carry out
factory inspections. This was considered an outrage by
most colonial employers accustomed to absolute discretion
in these matters. The proverbial last straw came when the
colonial governments instituted the Moyne recommendation
to assist in the development of the trade union movement.
Departments of Labor and Labor Exchanges were set up
throughout the Caribbean. Guyana's Department of Labor
went into operation in 1942.

The policy reversal was not however the betrayal some
employers were prone to assume. The intention of the
colonial administration was to subdue and educate the
labor movement emerging in its midst. Colonial bureaucrats and British trade unionists worked diligently to
develop desperately needed organizing and collective
bargaining skills among the emerging labor leadership but
their "dogma was absolute and complete cooperation and
collaboration between labor and capital."[25]   In short,
the spontaneous development of a cadre of militant labor
leaders was to be aborted and an elite group of labor
bureaucrats implanted in its place. Furthermore, although
the colonial government was ready to train and assist
labor bureaucrats, it still refused to use the legal and
coercive powers of the state to provide the permanent
security arrangements needed by trade unions. Employers
were left free to accept or reject a union as the bargaining agent for employees.

Under the British voluntarist tradition there are no
legally prescribed procedures for (1) determining union
membership strength, (2) resolving inter-union representational disputes and (3) compelling recognition of unions
with majority support. Thus the colonial government left
employers with more than an adequate arsenal to prevent
the development of a free trade union movement. Instead,
the voluntarist system of labor-management relations has
spawned sweetheart unions, a large number of small unions,
cannibalistic rivalry and a high frequency of strikes.
Thus, although workers made known their discontent and
willingness to organize by flocking to the trade unions
which sprang up during the 1940s and 1950s a strong TUM
did not develop "... finances were very slender and
unions, not even the oldest established one, could afford
to provide the necessary services of a modestly run trade
union."[26]

The Modernization of Bookers

As the British Colonial Administration busily laid
the groundwork for an acceptable decolonization, Booker
McConnell, Ltd., Guyana's largest colonial interest, also
prepared for the inevitable. Booker McConnell, Ltd.,
better known as Bookers, was Guyana's Sugar King. Josias
Booker I first came to Guyana in 1815. By the end of
WWII, the colony's economy and the colonial state virtually belonged to the company. With more truth than
jest, inhabitants referred to their country as "Booker's
Guyana," and claimed that the company's far flung businesses could do everything except bury you. The exag-
geration was only slight. Bookers controlled 70 percent
of the colony's sugar production. Booker's ships carried
the colony's trans-Atlantic trade. The stevedoring,
warehousing and waifing businesses belonged to Bookers.
The petroleum market company, a taxi fleet, a printing and
packaging firm, a foundry, a drug company, an insurance
company, and the colony's largest wholesale and retail
businesses all belonged to Bookers. Bookers had even
found profitable business opportunities in the hinterland
bleeding balata trees. The scale and diversity of Bookers' operations in the local economy were only matched by
the population's hatred for the economic colossus.

The story of Bookers' post-war transformation is
amazing and exemplory of the far-reaching changes which
have occurred in the world capitalist system over the last
forty years. At the close of WWII, Bookers vast economic
holdings were solidly rooted in the Caribbean sugar
industry. While Bookers' non-sugar businesses were
extensive and expanding, sugar profits were the heart beat
of the company. Nonetheless, with remarkable foresight
Bookers' management successfully pursued a corporate
strategy which transformed the company into a major
multinational corporation with only negligible interests
in tropical agriculture. Today, the corporation seeks to
completely divest itself of equity involvement in the
Third World in favor of less risky, contractual relations.
The architect and driving force behind this feat of
corporate adaption was Lord Jock Campbell.

Jock Cambell joined Bookers' Board in 1945 and became
its chairman in 1953. He subsequently earned the distinction of becoming a living legend within the company.
Bookers' current Chairman, Chief Executive Officer,
Company Secretary and other members of senior management
speak of Lord Campbell with a respect that borders on
veneration. He is recalled as a man of great vision.
Although we sought to interview Lord Campbell, he was

unavailable. Nonetheless, the charisma of this man who master-minded the transformation of Bookers' corporate image and policies, was felt at every interview in London and several in Guyana as well.

Jock Campbell was the heir to a fortune built on five generations of sugar estate ownership in British Guyana. Yet when he joined Bookers' he brought with him an advanced understanding of post-war political and economic trends. Jock felt that de-colonization was both inevitable and imminent. Moreover, he firmly believed that colonial companies would have to completely transform their modus operandi if they hoped to survive. Needless to say, such ideas were considered scandalous among the exploiting colonial classes. This charge was buttressed in Campbell's case by his public support for the British Labor Party. Ideas such as the following did not make Campbell a popular character in London's exclusive clubs:

> The governments of new nations, in the face of formidable social and economic problems, have no straightforward choice between capitalist and communist systems of economic development... Strong central planning and control by their governments seem essential to the progress of most of these countries...It is Western capitalism which must if it is to survive, adapt itself to, and become part of, the historical processes of these nations.[27]

In addition, Campbell warned that the world-wide, anti-colonial movement would not be satisfied with mere political independence. With uncanny accuracy he predicted the current struggle for a New International Economic Order.

> In Central Africa there is the struggle between the understandable and inexorable demand of Africans for majority government, and the romantic illusion of some Europeans that they can continue to dominate the political scene... I am convinced that broad and lasting prosperity can only flow from narrowing the gap between rich and poor countries...What is wanted is a Declaration of Interdependence.[28]

Despite the strength of his convictions and his unusual leadership skills, Campbell ran into resistance from Bookers' other board members and senior management. He had a particularly tough time convincing Bookers' top management in Guyana. These men were convinced of the natural inferiority of the natives and their God-given right and duty to rule backward people. Mr. Seaford, one time chairman of Bookers' Sugar Estates (BSE) is a prime example of this colonizer mentality. In 1939, he unabashedly told the Moyne Commission that the Indian laborer was carefree, foolish and easily satisfied with the barest necessities of life--a little food, bright colored clothing and the opportunity for easy love-making! Needless to say, Campbell's program for the modernization of Bookers' productive techniques and social relations represented a challenge to everything old-fashioned managers like Seaford stood for. Many of them had risen through the ranks and lacked any formal training. Their management skills left much to be desired. According to Brian Scott, their style "was often demanding, paternalistic, and old-fashioned, and of a low intellectual quality."[29] As a result, the company had a reputation for ruthlessly extracting its profits from the blood of sugar workers who were sympathetically regarded as little more than slaves by the local population. Campbell was determined to change this. Belately, Guyana's sugar industry was to be brought into the 20th century through a combination of modern capitalist efficiency and a new philosophy of corporate responsibility.

When Campbell visited Guyana right after the war, he found Bookers' holdings in a deplorable state. Cane fields had not been properly maintained. Factories and equipment were archaic and run down. In 1947, Bookers' chairman, described the challenge facing Bookers in these words:

> Since 1939 it has not been possible to obtain more than the most meagre supplies of the machinery, equipment and spares needed for replacements, renewals and repairs in field and factory on the sugar estates...And we are now faced with an enormous rehabilitation program, with supplies still short, and costs, which in many cases have already doubled still mounting.[30]

Campbell envisioned this "enormous rehabilitation program" as only one aspect of the Bookers' modernization. Mechanized field operations and modern factories would be meaningless, he argued, without greatly improved industrial and community relations. Starvation wages would have to be increased, labor-management relations derived from slavery would have to be brought into the 20th century and the social isolation of the plantation would have to be dissolved. Campbell's two pronged approach to Bookers' corporate transformation is exemplified in a letter he wrote to the BSE Board in 1946:

I'm preaching that fact that...we cannot distri-
bute profits at the expense of assets--human or
material. Proper living conditions for staff and
labor must be a charge on the business parallel
with the maintenance of our land and equipment in
the highest economic state. There's no doubt
that of late years many of our estates have shown
profits at the expense of the future. Admitted
some of these profits were put to reserve and
are available in the last resort, but how far
better would our outlook be if surplus profits
had been spent on essential housing and equipment...The Board need not fear that I'm
encouraging expenditures on luxuries--far from
it. The point is that certain things must be put
right if the industry is to survive--factories
must be efficient, staff and nuclear labor must
be well housed and cared for, mechanical field
equipment must be really good, our technical
staff must be strengthened--all these are minimum
requirements.[31]

Campbell was emphatic. A colonial company could no
longer prosper without concern for the development of the
colonial territory or its people. In the second half of
the 20th century, modern capitalist efficiency would have
to replace the colonial tradition of brutal labor exploitation as the basis of profit in the sugar industry.

Campbell was determined to transform Bookers into the
most efficient sugar producer in the Caribbean "and even
in the British Empire from all points of view--field,
factory, staff and labor."[32] The physical and technical
aspects of the program went ahead with great speed and
success. The financial backbone of Bookers' post-war
policy was provided by the Commonwealth Sugar Agreement
(CSA). CSA was reached in 1949 and guaranteed stable and
relatively high sugar prices for at least the next eight
years. Thus, after over a century of capital starvation,
investment funds were readily available.[33] Millions of
pounds were spent to expand acreage, mechanize field
operations and modernize factories. The program was a
remarkable success. The increase of acreage and productivity is reflected in the rapid increase in the colony's
sugar output. In 1945 Guyana produced 157,000 tons of
sugar. By 1952, the figure had soared to 242,692.[34]

Bookers' initiative was also very successful in
respect to the reorganization of the company's diverse
businesses in Guyana. Throughout its long history in
Guyana, Bookers had grown without rational planning.

Bookers was made up of a broad range of businesses
including such diverse and unrelated operations as balata
collection in the interior and hotel management in Georgetown. In Campbell's words the company was a "shapeless,
incoherent conglomeration of variegated activities."[35]
To be more accurate, this critical description was true
for only Bookers' non-sugar business, i.e., drug manufacturing, petroleum marketing, taxi service, etc. In
regard to sugar and its by-products, Bookers' companies
formed a vertically well integrated operation. BSE
cultivated the cane and manufactured the sugar. Demerara
Sugar Terminals provided for storage and handling.
Bookers Shipping owned the wharf and provided stevedoring
and ship services. The Booker Line supplied the vessels
which made the transatlantic voyages. Finally, Bookers
Sugar Company and United Rum Merchants handled the
marketing and distribution in the industrialized countries.

Bookers' organizational problems stemmed from the
predominance of sugar over the non-sugar businesses. The
chairman explained the needed reorganization as follows:

> We have considerable administrative problems to
> solve in our British Guiana undertakings because
> their organization has in the past been almost
> wholly oriented towards sugar. As part of our
> reorganization scheme, we intend to ensure that
> our shops in British Guiana shall have the admin-
> istrative attention--and thus the administrative
> efficiency--which their importance justifies.[36]

In 1951, Bookers' varigated operations were divided
into five separate groups of companies each with its own
Board. The explicit purpose of this grouping of companies
was to increase operational efficiency. An unintended,
but far-reaching consequence of this division of companies
was its contribution to a new public perception of Bookers. Bookers, up to this point, had been perceived as a
monolithic structure--the whole personality of which was
determined by sugar. The break-up into groups of companies, many with no relation at all to sugar, helped to
undermine this corporate image. Gradually, Guyanese came
to see Bookers as being composed of large and small
companies, profitable and marginal businesses and, most
importantly from the point of racial perception, as the
estates versus the Georgetown Group of companies.

Not surprisingly, Campbell's initial efforts to re-
volutionize Bookers' social relations were not as successful as the company's technical and organizational

innovations. Bookers management in Guyana remained
staunchly opposed to what was perceived as Campbell's
screwball ideas of social improvement and equality of
opportunity. In 1946, the British Guyana Sugar Producers'
Association (BGSPA) appointed its first social welfare
officer primarily to organize recreational facilities for
the estates' labor force. The following year the Sugar
Industry Labor Welfare Fund went into operation and monies
became available to improve the sub-human standard of
housing facilities for sugar workers. However, despite
the overwhelming needs of the population on the estates,
the scale of Bookers' Social Welfare and Housing Programs
remained meagre.

Indeed, the Board of Bookers Sugar Estates referred
to this aspect of Campbell's modernization program as
"bread and circuses."[37]   A major political crisis
finally convinced the old-timers that Campbell was
right.

Emergence of Guyana's Socialist Oriented Independence Movement

As previously noted, the liberalization of British
colonial policy was not born in a political vacuum. Colonial subjects everywhere including the Caribbean were
throwing off habitual docility in favor of mass defiance.
Although there was undoubtedly some genuine concern for
improving the condition of workers involved in the re-
formed labor policy of colonial governments, the more
urgent objective was "...to isolate the trade unions from
the political hustings. There was almost an evangelical
proselytizing of the thesis that trade unions should
remain aloof from politics."[38]   The unity of political
and trade union issues and the control of the trade unions
by middle class politicians dates back to the beginning of
the labor movement in the Caribbean. On the one hand, the
rank and file believed their cause would benefit by close
association with middle class leadership. On the other
hand, aspiring politicians convincingly argued that
improvements in wages and benefits could only be won after
sweeping constitutional and political reforms. Political
democracy--universal adult suffrage, internal self-
government and ultimately independence--became the rally-
ing cry of the Caribbean labor movement. The Jamaican
elections of 1944, the first in the region to be held
under universal suffrage, "...made it abundantly clear
that workers' votes would be cast solidly in support of
their trade-union sponsored candidates regardless of

political ideology or the absence thereof." Little
political acumen was needed to see that state power
would "devolve upon the most powerful trade-union complex."[46]

Guyana's political and trade union development fit
the Caribbean pattern quite well. According to Aston
Chase, "...nearly everyone of the politicians of substance
in the late 1940s and throughout the 1950s was at one time
or another connected with the trade union movement."[40]
In the immediate post-war climate of uncertainty and
impending political change, the future leaders of the PPP
evolved a two-pronged strategy to transform the democratic
agitation of the preceeding decade into a full-fledged,
mass-based independence movement. In 1946, a group of
political activists got together and formed the Political
Affairs Committee (PAC). Up to now Guyana's colonial
system had carefully restricted political participation to
a tiny fraction of the indigenious population. For
instance, the new Constitution of 1945 still imposed
prohibitive property and/or income qualifications on the
right to vote. As a result, the Black and East Indian
middle class found itself systematically excluded from
inner circle of political power despite their new economic
privilege and social status. PAC provided a structure
wherein the members of the rapidly growing middle class
could come together to discuss competing political phil-
osophies and practical plans for achieving national
independence. Organized with the help of the Communist
Party of Great Britain,[41] PAC displayed a clear bias
towards more radical interpretations of the colony's
situation. In fact, the founding members of PAC, Cheddi
Jagan and his wife Janet, Jocelyn Hubbard and Aston Chase,
were deeply influenced by marxism-leninism. Despite
Guyana's underdevelopment, many found the marxist model
quite attractive. Seeking to organize a broad-based
nationalist movement in a racially and culturally divided
society, many members of PAC were reassured by the marxist
assertion that racial conflict is epiphenomenal and a
capitalist ploy to distract attention away from the real
issue of class exploitation. Believing this to be true,
they felt they could avoid divisive racialism and focus
their movement on the common element of exploitation by
foreign capital.

PAC drew great attention among the intelligentsia and
became the hub of political activity in the capital city
of Georgetown. Its publication, the PAC Bulletin, featured marxist interpretations of Guyanese history,
current events and news reports from communist countries.
It was widely read by educated Guyanese. However, among

the poor, PAC had little direct impact. Largely un-
educated, desperately poor and consumed by the daily
struggle for survival, the common Guyanese worker had
little time or interest in debates about abstract economic
and political issues. It was obvious that PAC would have
to address the immediate needs of Indian and Black workers
if it hoped to mobilize the masses in an anti-colonial
movement. Inspired by the success of other Caribbean
nationalist leaders in building a unified political-trade
union complex, PAC's key members devised a strategy to win
control over the Guyanese labor movement. In this effort,
they were assisted by the Caribbean Labor Congress (CLC),
a regional organization of nationalist and trade union
leaders in which Caribbean marxists played a leading role.
Moreover, through the CLC the radicals within PAC established ties with the World Federation of Trade Unions
(WFTU) and the international communist movement.

The linchpin of the PAC's trade union strategy was to
capture control of the Manpower Citizens' Association
(MCPA) based in the sugar industry. In comparison to the
other unions, the MPCA was "a giant among pygmies....All
other industries and even the Government paled in comparison with sugar as to the numbers of persons employed
by them."[42] With the largest trade union in the country
under its control, PAC would easily, it speculated, be
able to bring the National Trade Council, the umbrella
group for organized labor, under its command. The plan
was elegant in its simplicity and the MCPA, as a dis-
credited company union, was ripe for takeover. By 1946
its membership had plummeted from over 20,000 to a mere
600 members. The MCPA's newsletter, "The Labor Advocate,"
read like a promotional newsletter for the sugar companies
which supported the publication through heavy advertisement subscriptions. With MCPA officials on the BGSPA
payroll, the union had become a rubber stamp for the
companies' production and wage decisions. It was on this
breach of faith between the MCPA and the field workers
that the leaders of the PAC sought to build a bridge to
political power. In 1945 Cheddi Jagan became the
treasurer of the MPCA and the TUC.

The results of the 1947 elections confirmed the
fundamental assumptions underlying the PAC's political
strategy. Jagan won a seat on the Legislative Council
largely because of his trade union activity. Furthermore,
the inability of the sweetheart union to get more than one
of its candidates elected thoroughly exposed the bankrupt-
cy of the MCPA and suggested the workers' readiness to
support an alternative organization. Thus "infiltration"
of the TUM, preferable at the leadership level, became the

cornerstone of the PAC's political strategy.[43] Nonetheless, Jagan could not defeat the corrupt leadership
which received support from the powerful Sugar Producers'
Association, the conservative unions within the TUC
and the British Trade Union Council. Thus in 1947, Jagan
joined with Dr. J.P. Lachmansingh and A.A. Rangela, both
prominent communal leaders of the British Guyana East
Indian Association, to organize the Guyana Industrial
Workers' Union (GIWU). With the support of the majority
of sugar workers it was assumed that the GIWU would
successfully challenge the status of the MCPA as the sole
bargaining agent for field workers, win official recognition and become the most powerful trade union in the
colony.

Clearly, racial loyalty and symbolism played a
significant part in the mobilizing strategy of the
PAC/PPP from the very beginning. Lachmansingh and Rangela
became president and vice president respectively of the
new union. And, of course, Jagan realized that being an
Indian and coming from a long line of sugar workers made
him an attractive political leader in the sugar belt.
Moreover, PAC's marxist leaders were 'keenly aware that
they needed a Black leader of national prominence to
attract the large Black urban population. According to
Despres, this urgent need led the radicals to accept
L.F.S. Burnham into their inner circle despite their
serious reservations about his personal and political
character.[44]

To many within PAC, Burnham appeared as an ambitious,
petty bourgeois politician espousing a vague socialist
ideology. The heart of the problem from the perspective
of PAC's hard-core leadership was the fact that Burnham
was not a marxist and had eschewed all contacts with the
Communist Party of Great Britian while studying in London.
Nevertheless, upon his return to Guyana, Burnham became
the major Black spokeman for immediate independence and
president of the British Guyana Labor Union. Thus in
pursuit of its goal of a racially balanced leadership,
Burnham was elected chairman of the newly created Peoples'
Progressive Party and became the de facto leader of the
moderate socialists and non-ideological nationalists
within the developing anti-colonial movement. Cheddi
Jagan, the leader of the marxist faction within PAC, was
named second vice-chairman of the party. The rest of the
leadership positions were carefully divided between East
Indians, Blacks, Chinese and whites.

In sum, both personality and ideological differences
were submerged within the leadership of the new independence party. The basis for the coalition between marxists, moderate socialists, nationalists, Indians and

Blacks was a consensus supporting the development of a united, nationalist movement and the demand for immediate independence.

Needless to say, there were serious disagreements over how these goals could best be achieved and what to do after independence. From the outset the revolutionary rhetoric and bellicose style of the radicals caused fissures in the coalition. The marxist-leninist faction led by Jagan insisted that Guyana's struggle for independence be waged as part of the international, anti-imperialist struggle and therefore allied with the world communist movement.

According to Jagan the Soviet Union--not Western capitalism--offered the best model of how to rapidly develop a backward country. To popularize this revolutionary vision, the radicals imported at least half a million books and pamphets into the colony extolling the experience of existing socialist societies and explaining basic marxist principles. In addition, they cultivated their contacts with communists within regional and international organizations.

Not surprisingly, many of the moderate and middle class members of the PPP were embarrassed, irritated or alarmed by the actions of the radicals. Those who were primarily nationalists and envisioned local capitalist development were obviously alienated by the revoluationary line of the marxists. However, even members who more or less accepted the radical's interpretation of Guyana's situation "...considered that the actions of the extremists were stupid and ill informed."[45] In brief, the moderates accepted the British policy of decolonizaton wherein it was assumed that nationalists leaders would make radical gestures to win the support of the masses but once in power play by the established rules of the game. According to R.T. Smith, "This is the royal road to political independence and it could be argued that any sensible politician would take it no matter what his ultimate ends might be."[46] We will go a step further and suggest that the bellicose style and hard-line rhetoric of the PPP radicals were largely due to their superficial understanding of marxist theory and limited knowledge of the real history of the Soviet Union.

Between 1949 and 1951 the movement lost its more conservative members. The impact of this initial fissure was minor. Few individuals dissociated themselves from the party and those that did lacked links to important constituencies. It did, however, portend grave dangers to the future unity of the anti-colonial struggle. Most of the those who left were Colored and Black professionals and civil servants. They asked Mr. Burnham to leave with

them and organize a moderate alternative to the communist-
inspired PPP. For the time being, Burnham refused. As he
read the situation, the Indian vote would definitely stay
with Jagan and, with the support of the Black radicals,
Jagan could also count on significant support from poor
Black Guyanese.[47]   Despite his   decision to stay, the
lines of ideological and tactical dispute were growing
sharper and, most unfortunately, tended to coincide with
racial divisions. Many Black and Colored activists,
perhaps because of their more thorough assimilation
of British culture and education, tended to favor
a more accommodating approach to achieving independence
and economic development. Conversely, most of the Indians, who until quite recently were excluded from the
educated middle class and denied social acceptance,
rallied behind the fiery rhetoric of the radicals.

External developments simultaneously undercut the
continued unity of the Guyanese nationalist movement. The
emergence of the Cold War and its fall-out in the international labor movement could only exascerbate the existing division within the fledgling party. To a great
extent the outcome of the power struggle between the
radical and moderate factions of the PPP's leadership was
determined by the strength of their respective allies in
the international arena. The marxists nurtured their ties
to the Soviet Union and the World Federation of Trade
Unions. The moderates increasingly turned to the West and
joined the rabidly anti-communist International Confeder-
ation of Free Trade Unions.

The radicals developed the GIWU into the most powerful trade union in the colony but nonetheless failed to
win official recognition from either the Sugar Producers
Association or the National Trade Union Council. Consequently the only formal recognition the GIWU enjoyed
came from outside the country within the Caribbean Labor
Congress and the World Federation of Trade Unions, both of
which were strongly identified with the world communist
movement. When the power and influence of the CLC and
WFTU was destroyed by the ICFTU and the Inter-American
Regional Organization of Workers (ORIT), the radicals
found themselves isolated from the Caribbean nationalist
and trade union movement and without a legal base within
the Guyanese labor movement. In sum, the policy of the
U.S. State Department and the conservative wing of the
American labor movement aimed at purging "communists" from
all social movements in the Western Hemisphere and building pro-American trade unions and nationalist parties was
working extremely well.

Obviously, the non-marxists within the independence movement were not adversely affected by these developments. In fact, Despres argues that Burnham's position was considerably strengthened by the rising tide of anti-communist hysteria. When Grantley Adams of Barbados denounced the CLC as a communist front and Norman Manley expelled the marxists from the Peoples' National Party of Jamaica, Jagan's faction lost its key links to two of the most important nationalist movements in the British Caribbean. Burnham, on the other hand, continued to enjoy a close relationship with both men and their organizations. Moreover, the British Guyana Labor Union of which Burnham was president chose to join the ICFTU in 1952. The marxists were clearly not succeeding in their bid to assume the leadership of the labor movement. In early 1952 the GIWU replaced the MPCA on the TUC but by the year's end the national organization applied for membership in the ICFTU. The TUC was rejected because of communist infiltration, in short, because of the GIWU. The handwriting was on the wall. Under increasing pressures, the radicals and the moderates maintained their fragile alliance and went into the 1953 elections as a united front.

The elections were the first held under universal adult suffrage. No one, not even the party leadership, had expected the PPP to win. In fact, when the Waddington Commission visited British Guyana in 1950 and drafted the major provisions of the 1953 Constitution its five members predicted that no party would emerge capable of winning a legislative majority. Heretofore, numerous political parties had been hatched on the eve of elections for the exclusive purpose of mobilizing support for ambitious individuals who just as quickly dissociated themselves from any collective responsibility once in office. The outcome of the elections therefore came as a complete shock to the colonial establishment. The PPP devastated the opposition.

During the campaign the National Democratic Party, the political creation of the MPCA and the SPA, lambasted the PPP as communists. With money from the SPA, the NDP placed full-page ads in the newspapers warning that the "reds" in the PPP were agents of the Soviet Union seeking to impose a totalitarian state on the unsuspecting Guyanese people. The Undesirable Publications Act which prohibited the PPP's importation of "subversive" literature, authored by Lionel Luckoo, the chairman of the NDP and president of the MPCA, became a major campaign issue. Vitriolic speeches and name-calling substituted for any substantive discussion of the serious problems confronting

Guyana. The PPP countered the opposition's red-baiting
with condemnations of "the lackeys" serving the sugar
barons and "imperialist expoitation." Jagan concentrated
on Indian voters and Burnham on the Black voters without
making explicitly racial appeals. And, in a few districts
the party chose to run a candidiate who was not of the
same racial and cultural background as his constituency.
Despite the fiery rhetoric, the campaign was totally
peaceful and the PPP insisted that their victory would be
legal.

Finally given the opportunity to select their Government, the majority of the people overwhelmingly rejected
the representatives of the plantocracy. The PPP won 51%
of the vote and 18 out of 24 legislative seats. The NDP
won only 2 seats; moreover, both the president and secretary of the MPCA lost their electoral bids. The remaining
4 seats went to independent candidates. The Colonial
Governor had no alternative but to reluctantly ask the PPP
to form the new Government.

This decisive victory ironically confronted the
party with a major problem. In fact, the leadership even
debated the possibility of refusing to take office. They
had prepared themselves to play the part of the uncompromising opposition. As the gadfly they had intended to
use every action of the government as an opportunity to
expose the limitations of the paternalistic constitution
and thus push their demand for total and immediate independence. Ultimately the decision was made to assume
office and it would appear that the radicals' view pre-
vailed. The objective would be to embarrass the colonial
authorities by pushing the "real government" to the wall
to precipitate a constitutional crisis. Thus to understand the subsequent developments, R.T. Smith reminds us
that "They did not feel themselves to be the government of
the country . . . and much of the action of the elected
members between April and October of 1953, must be viewed
in the light of these facts."[48]

The PPP was in office for only 133 days when British
gunboats landed in the night to suspend the constitution
and remove the elected government. During this brief
interim of popular government, the PPP enacted several
pieces of progressive legislations and ensured that
representatives of working people were given "pride of
place" on governmental committees.[49] At no time however
did the PPP Government violate the constitution or under-
take any illegal action. As Despres observed, "If the
government was guilty of anything, it was guilty only of
demonstrating that Guyana remained a colony of the British
Crown."[50] Nevertheless, the religious establishment and

the Sugar Producers Association declared that an illegal communist takeover was underway. The "crimes" of the PPP were firstly its attempt to end sectarian control of publicly financed schools and, most importantly, its support of a Labor Relations Bill which would have entitled workers to select their union representative by secret ballot.

In September 1953 the GIWU once again went on strike to demand recognition as the legitimate bargaining agent for workers in the sugar industry. With its party in power and the overwhelming support of the workforce it had no doubt that this time it would oust the MPCA. The PPP, deeply desiring to see its major foothold in the trade union movement gain official status, mobilized its resources in behalf of th GIWU. According to the report by the British commission sent to investigate the cause of the constitutional crisis:

> Several PPP ministers (notably Mr. King) who, as members of the Executive Council, were responsible for peace and order and economic development of the country, together with many PPP members of the House of Assembly, toured the sugar estates, calling upon the workpeople to support the strike and using language which was bound to encourage . . . acts of intimidation and of violence against those who refused to observe the strike call . . . Altogether the strike lasted 23 days and in that period strenuous efforts were made by the leaders of the PPP, including several Ministers, to get the strike extended into a general strike covering the whole country.[51]

The strike was abruptly called off when the PPP announced that it would submit legislation to the House which would protect the workers' right to democratically elect their union representative. Since the PPP had an absolute majority in the Assembly passage was thought assured.

Many commentators have argued that the PPP was only seeking to enhance its own position within the labor movement by pushing the Labor Relations Bill of 1953. For example, Despres points out that under the proposed bill the Minister of Labor (obviously a leader of the PPP) would have had complete authority to oversee the union elections.[52] In a similar view, Mr. Romualdi, of the ICFTU and ORIT, charged that Government was trying to force the SPA to break its contract with the MPCA and "to collar the workers into a government union."[53]

Obviously the interest of the PPP would have been served by the Labor Relations Bill, which, by the way, was more moderate than the American Labor Relations Act on which it was based. The MPCA could not win a free and fair election within the sugar industry and the GIWU would have become the recognized union for the single largest group of wage earners in the country. To conclude from this, however, that the PPP was only seeking the gain control of the TUM or reward Indian voters, in particular sugar workers, is to miss the class-wide significance of the PPP's labor policy. Had the legislation taken effect the foundation for an alliance between workers and a socialist-oriented Government would have been firmly laid. The British voluntarist tradition had not only protected a renegade union in the sugar industry. More generally, it had dissipated the vitality of the entire TUM.

Our point is simple. During the early phase of the PPP's history both its ideological posture and practical considerations lead the leadership to focus upon the creation of a strong trade union movement. Surely, the PPP recognized that collective identities grounded in racial and ethnic differences had to be taken into consideration; but, the lynchpin of its early strategy of mass mobilization was the class struggle and class solidarity. Even the moderate wing of the party agreed that a powerful labor movement was necessary to wrest state power from the colonial authorities. It was Forbes Burnham who introduced the Labor Relations Bill to the Assembly in 1953. The PPP was assured that with minimal legislative support, a strong and united TUM would develop; and , more importantly, that such a movement would support its leadership. The British agreed with this prognosis and acted decisively to prevent any such development. The gunboat suspension of the constitution was only the first step.

NOTES

1. Harold Lutchman, "Perspectives on the Role of the
    Trade Union Movement in Guyana," Release, 6 and 7
    (1979).
2. Walter Rodney, A History of the Guyanese Working
    People, 1881-1905. (Baltimore and London: The Johns
    Hopkins University Press, 1981), p. 33.
3. Ibid., p. 34.
4. Orlando Paterson, "Context and Choice in Ethnic Alle-
    giance: A Theoretical Framework and Caribbean Case
    Study," in Ethnicity: Theory and Experience, ed. N.
    Glazer and D.P. Moynihan, (Cambridge, Massachussets:
    Harvard University Press, 1975).
5. Ibid., p. 343.
6. Rodney, p. 102.
7. Ibid., p. 183.
8. Leo Despres, Cultural Pluralism and Nationalist
    Politics in British Guyana. (Chicago: Rand McNally
    and Company, 1967), p. 50.
9. Ibid., p. 52.
10. Harold Wolpe, "Theory of Internal Colonialism: the
    South African Case," in Beyond the Sociology of
    Development, edited by Ival Oxaal, Tony Barnett and
    David Booth, (London: Routledge and Degan Paul,
    1975), p. 244.
11. Rodney, p. 218.
12. Ibid., p. 187.
13. Ibid., p. 181.
14. Ibid., p. 102 and p. 210.
15. Ibid., p. 22.
16. Aston Chase, A History of Trade Unionism in Guyana,
    1900-1961. (Ruimveldt, Demerara, Guyana: New Guyana
    Company Ltd., n.d.), p. 27.
17. Ibid., p. 13.
18. Ibid., p. 33.
19. Ibid., p. 20.
20. R.T. Smith, British Guyana. (London: Oxford Univer-
    sity Press, 1962), p. 146.
21. Ibid.
22. Philip Reno, "The Ordeal of British Guyana," Monthly
    Review, 16 (July-August 1964): 93-94.
23. Zin Henry, Labor Relations and Industrial Conflict in
    Commonwealth Caribbean Countries (Trinidad: Columbus
    Publishers, Ltd., 1972), p. 49.

24. Brian Scott, "Organizational Network: A Strategy Perspective for Development" (Ph.D. dissertation, Harvard University, 1979), p. 116.
25. Chase, p. 104.
26. Ibid., p. 121.
27. Booker McConnell, Ltd., Annual Report and Accounts 1959, Statement by the Chairman.
28. Booker McConnell, Ltd., Annual Report and Accounts 1961, Statement by the Chairman.
29. Brian Scott, "The Organizational Network: A Strategy Perspective for Development" (Ph.D. dissertation, Harvard University, 1979), p. 116.
30. Booker McConnell, Ltd., Annual Report and Accounts 1946, Statement by the Chairman.
31. Quoted in Bookers' Sugar, supplement to Booker McConnell, Ltd., Annual Report and Accounts 1954, p. 89, original emphasis.
32. Ibid.
33. Scott, p. 116.
34. Guyana Sugar Corporation, "The Guyana Sugar Industry, Production and Exports," Company files, GUYSUCO, Georgetown, Guyana.
35. Quoted in Scott, p. 125.
36. Booker McConnell, Ltd., Annual Report and Accounts 1949, Statement by the Chairman.
37. Quoted in Scott, p. 158.
38. Chase., p. 104.
39. Henry, p. 32.
40. Chase, p. 104.
41. Despres, p. 184.
42. Chase, p. 85.
43. Ralph Premdas, "The Rise of the First Mass-Based Multi-Racial Party in Guyana," Caribbean Quarterly 20 (September-December 1974): 5-20.
44. Despres, p. 188.
45. Smith, p. 178.
46. Ibid., p. 173.
47. Ibid., pp. 178-179.
48. Ibid., p. 173.
49. Chase, p. 202.
50. Despres, p. 209.
51. Quoted in Bookers' Sugar, supplement to Booker McConnell, Ltd., Annual Report and Accounts, 1954, p.65.
52. Despres, p. 208.
53. Ronald Radosh, American Labor and U.S. Foreign Policy. (New York: Random House, 1969), p. 399.

CHAPTER 3

Bookers' Guyanization and Imperialist Intervention

The Sugar Producers' Association had been shocked by the PPP electoral victory. In light of the PPP's actions in office, the SPA viewed the situation in desperate terms: "The colony was now in the hands of a Government bent not upon construction but destruction and one, moreover, prepared to descend to any means to achieve its ends."[1] Needless to say, Bookers, the leading member of the SPA, was greatly relieved when British troops arrived to suspend constitutional government in the colony. This time Bookers' interest had been protected by coercive measures. The more astute in the business community and the Colonial Office realized that such could not be the case in the future. Under the circumstances, Campbell's "utopian" ideas sounded more and more reasonable to Bookers' local Board members. Although Bookers threatened Guyanese that the company would pull up stakes and abandon the colony to economic ruin, this was not a realistic option. Bookers' fortunes were inextricably tied to Guyana.

> In 1950, 50% of the total assets of the company were located in the Caribbean, and in the period 1951-55, 66% of the profits also came from this area. At that time Bookers' Caribbean investments outside Guyana were minimal. Furthermore, a large part of the companies other assets and activities, such as shipping, sugar marketing, supplies procurement, and the spirits business, were to a considerable extent, reliant on Guyana. Bookers was, therefore, very much bound up with the fortunes of the colony whether it liked it or not. The company was virtually forced to remain in Guyana and search for practicable solutions to its problems.[2]

In light of these circumstances, most of Bookers' management could now at least agree that the company's image as a rapacious colonial exploiter had to be changed. "Bookers' Guiana" had to be transformed into Guyana's Bookers, a company identified with the welfare and aspirations of all Guyanese.

The processes by which Bookers sought to accomplish this extraordinary reversal of public opinion is referred to as Guyanization. In a letter to Bookers' top management in Guyana, Campbell outlined the main objectives:

. . . a "foreign" industrial undertaking imposed
upon the life of a people cannot gain their
confidence--and ward off their hostility--however
good conditions of service unless a sense of the
undertaking belonging to the country and the
people can be achieved; unless, above all, the
people can be brought to understand their identity of interest with the undertaking--and their
stake in its success . . . In the last resort
confidence, and a sense of belonging and of
identity of interest, can only be achieved by
real integration of the life of the industry with
the life of the community--especially by jobs at
all levels being available to the people of the
country, and by the companies' employees taking
part in the social, economic and political life
of the community.[3]

In sum, an old British expatriate firm was to be transformed into a Guyanese institution--at least in appearance. Obviously, no one really knew just how this was to
be accomplished in 1953.

Bookers experimented with a number of different programs. The unsuccessful social welfare program started in
1946 was injected with new funds and personnel. By 1956,
social welfare officers were busy on every estate trying
to develop community councils, tenant associations,
educational councils, youth clubs, libraries, film shows,
cricket, athletics, and handicraft clubs. Bookers even
tried to develop cooperative societies on the estates. In
1953, the Board resolved to explore the possibility of
local share ownership. By 1960 action resulted. Five
hundred-eighty-two employees, members of the public and
local institutions bought shares in the newly created
Guyana Industrial and Commercial Investments Co. Furthermore, Bookers' Chairman pledged that "Wherever and whenever suitable, it is our intention to widen the opportunities for employees and other members of the local public
to hold shares in our overseas companies.[4] A decade
later, Bookers Store, the largest department store in the
country, became a public company with the issue of
G$400,000 worth of new shares.[5]

In 1956, BSE launched a pilot cane farming scheme
at the Wales Estate. The explicit purpose was to eventually turn over a portion of the country's cane cultivation to a class of small independent farmers loyal to
Bookers. In response to the failure of the pre-1953
housing program, Campbell came up with the idea of providing low-interest housing loans to sugar workers out

of the Sugar Industry Labor Welfare Fund (SILWF). With
these loans, workers would build homes on land leased from
BSE. The program was an immediate success and the new
housing areas grew rapidly. In response the the increasingly politicized issue of land hunger, Bookers
leased over 35,000 acres between 1956 and 1965 to redundant workers, cooperatives, the government and house
builders outside the estate areas.[6]

It was not, however, these social welfare programs
which had the most profound impact on Guyanese society.
Rather it was Bookers' vigorously pursued training and
affirmative action employment policy which quite literally
changed the face of Bookers and consequently the entire
society. Prior to 1953, it did not matter what type of
technical or professional training a Black or Indian had
managed to achieve. There simply was no place for him in
Bookers' senior staff. However, in the aftermath of the
constitutional suspension, Bookers' systematically recruited Black and brown men to fill positions which had
always been the exclusive preserve of expatriates, local
whites and Colored persons. It is difficult to convey the
social and psychological revolution this represented. The
colonial class structure which automatically determined a
person's life chances on the basis of his or her skin
color was shaken to its foundation. Bookers' new policy
was to remove white expatriates from the creolized racial
hierarchy and the disregard the local tradition which
placed Portuguese, Coloreds, Blacks and Indians in de-
scending order of social wealth, power and prestige.

In July, 1946, Jock Campbell had written a strongly
worded letter to Bookers' Boards in Guyana which unequivo-
cally stated that its companies were to pursue a nondis-
criminatory employment policy. He said in part ". . . The
policy of Bookers' Board is that there should be no discrimination of any sort or description whatever, against a
good man merely because he is colored. We need not
elaborate this as that is a clear statement of policy."[7]
Nonetheless, as in the case of Campbell's other social
policy innovations, little was done until after the emergence of the political crisis in October 1953. Then the
wheels began to turn and they turned quickly. Before the
end of 1953, Bookers announced it was seeking qualified
Guyanese of any color for senior positions. Bookers
eagerly courted Black and Indian professionals, senior
civil servants and noted sports stars to join the Bookers'
team. The political crisis had forced Bookers to jump
beyond passive nondiscrimination to actively pursue a
program of affirmative action. The 1958 Annual Report
announced the turnabout, ". . . it is our absolute policy,

resolutely pursued, not now to appoint anyone from outside
. . . to any job for which we can recruit a Guyanese . . .
with the necessary qualifications and experiences."
Furthermore, recognizing that the tradition of educational
and occupational discrimination had left most Black and
Indian Guyanese unprepared for such opportunity, Bookers
launched two major training programs in 1955. Between
1955 and 1975, Bookers spent over G$10,000,000 on its
cadetship and apprenticeship programs. When support for
on-the-job training and after-hours study assistance are
added on, the figure rises to G$20,000,000.[8]

The cadetship program was designed to identify young
Guyanese men and women with the secondary school training
and intellectual ability to complete a program which would
prepare them for the top managerial, technical and professional positions within Bookers. Every year, hundreds of
applicants were screened, but only a handful were chosen.
The standards were rigorous because Bookers hoped to
implement this aspect of Guyanization without any loss to
the company's efficiency. Applicants went through a
day-long battery of tests followed by extensive interviews. Only the few who performed with all around distinction were awarded the attractive scholarships to study
abroad at universities in the United Kingdom or the
Caribbean.

By 1966, the year independence was finally granted,
80 cadetships had been awarded for the study of agriculture, accountancy, chemistry, engineering, personnel
management, store management and pharmacy. Eighteen were
still in training and 25 had permanently left the country
or failed to complete the program. That left 37 successful graduates working in Bookers. Twenty-seven held
professional posts on the senior staff, nine held senior
administrative or commercial positions and one was a
director of two companies within the Bookers Group.[9]
Black and Indian members of the senior staff received the
same salaries as their expatriate counterparts with
equivalent training and experience.[10] In sum, the data
support Mr. Haynes, Bookers' current Executive Officer,
assertion that "Guyanization was not a local window-
dressing operation." Responsibility for the local operation of Bookers' businesses was handed over to local
personnel. In Mr. Haynes' words, "We became invisi-
ble."[11] In 1953, there were 350 expatriates; in 1960,
200; in 1969, 120; and in 1976, the year the sugar industry was nationalized, only 33 expatriates were still
employed by Bookers in Guyana.

In 1955, Bookers also started an apprenticeship
training program. The purpose of this program was to
provide craft and specialized skill training to manual

workers for Bookers' sugar estates and its rapidly growing
number of light industries. Not surprisingly, far more
people were reached by this program than the cadetship
program. By 1966, 391 apprentices had been selected; 150
skilled craftsman had graduated; and 170 were still in
training.[12] Obviously, the purpose of Bookers training
program was more extensive than the mere creation of a
group of local men to replace the privileged expatriate
stratum in Bookers. According to Harold Davis, Personnel
Officer for BSE in 1967, Guyanization was a systematic
effort to intensify the educational training of Bookers'
labor force and "to produce more skilled Guyanese at all
levels and in all aspects of industrial operations."[13]
Everyone in Bookers was encouraged to see the possibility
for advancement. In addition to the formal cadetship and
apprenticeship programs, there were many other informal
opportunities for employees to improve their skills and
education. For instance, in 1966 alone, over a thousand
individuals at supervisory grades received special training while 1,300 members of the middle and senior management also received training.

    Much of this intensive drive to upgrade the edu-
cational and skill level of Bookers' labor force and its
abrogation of traditional employment practices was neces-
sitated by the company's modernization of field and
factory operations. Bookers now needed men with university degrees and specialized technical training to oversee
production. Notwithstanding the above, there was also a
strong public relations component to Bookers training and
personnel policy. As a integral part of Campbell's
Guyanization program, it aimed, according to Bookers'
Human Relations Committee to make "Bookers in British
Guiana a Guianese enterprise in the fullest sense, a part
of the community and understood to be a part of the
community."[14] The extent of identification the new
members of Bookers' senior staff made to the company
was extraordinary in view of Bookers' historic reputation.
Twenty-years later and after the company's nationalization, many of these individuals--who now manage the
state-owned companies--still speak of Bookers in terms of
"we" and with obvious loyalty.

    With similar efficiency, Bookers stripped its George-
town management of the mystique and social aloofness which
had always characterized this powerful elite. This was
Tony Tasker's job as Bookers' first Public Relations
Officer. He was sent from London with instructions to
break down the traditionally impenetrable barriers between
the community and Bookers' elite. According to Mr.
Tasker, his job was to explain Bookers to the Government

and the community and to promote participation by Bookers'
top level people in all aspects of community affairs.
They were to make their expertise and experience freely
available to the Government and civic groups to promote
the country's all around development. Moreover, all of
Bookers' senior staff members were encouraged to socialize
with the native population.[15] One can only appreciate
what a profound change this represented when it is recalled that as late as 1956, senior staff members were
still required to obtain permission to leave the estates
even when off duty. The old expatriates and local whites
bitterly resisted this violation of the colonial taboo
against social mixing of the races. The young expatriates
became the toast of the town in the eyes of the Black and
Indian communities when they danced with Black and
Indian women in public. The cumulative impact of these
policies were precisely what Bookers was looking for:

> ...here the whole field of operations has become
> visible for the first time. People in Georgetown
> know what is happening in Bookers; it is no
> longer unusual to meet senior executives social-
> ly. It is no longer unusual to know, personally
> even a Director of Bookers.

Most importantly,

> Where before Bookers and sugar planters meant
> more or less the same thing in Georgetown as in
> the country, today Bookers for people in George-
> town mean mainly shops and industry and so on.
> The sugar aspect of the image has receded.[16]

As implied in the above quotation, not everyone was
equally impressed by Bookers' Guyanization program. Bookers' historic image as a loathesome employer and an evil,
omnipotent political force persisted in the Indian community despite Bookers' post-1953 face lift.[17] Conversely, for the Black population concentrated in the
urban areas, Bookers came to be seen as a good employer
and social crusader. The reasons for this divergence of
opinion between Blacks and Indians who were previously
united in their hatred of King Sugar are rooted in the
cultural, geographic and economic differences which
characterized the two racial groups.

The Black and East Indian Response to Bookers' Guyanization Program

As previously noted, Blacks and East Indians were introduced into the colonial economy under very different conditions and subsequently developed in different directions. Ex-slaves, after failing to establish themselves as an independent peasantry, developed a disdain for the land and all forms of agricultural labor. Instead, Blacks turned to the cities and mining towns where they satisfied the government's and new industries' need for cheap, unskilled labor. By 1960, Blacks constituted roughly half of Guyana's urban population. The dream of every Black family was for one of its children to achieve a secondary school education -- perhaps then he could win a government scholarship and go abroad to study law or medicine. Needless to say, precious few realized this dream of upward mobility. Nonetheless, a strong motivation towards academic achievement and a firm commitment to British values were cultivated in the Black community. By way of education and cultural orientation, this section of the local middle class was ready to fill the newly opened senior positions at Bookers.

Bookers' Guyanization programs suffered from a definite urban bias and according to Lloyd Searwear, who served a consultant to the company at the time, this represented   its major drawback. The new expatriates selected by Jock Campbell to implement his reformist policies concentrated their energies on Georgetown and Bookers' companies there. As a result, Blacks concentrated in the urban areas, particularly Georgetown, saw the best side of Bookers' modernization program. Members of the Black educated elite eagerly responded to the company's recruitment of local personnel to fill top management and technical positions. In a less dramatic manner, the Black urban working class saw new opportunities open up as Bookers' shops increased the scale of their operations and the parent company expanded into numerous new light industries.

John Huddart, a labor relations specialist, was hired by Bookers after it dropped its historic opposition to trade unionism to smooth the transition to a modern system of labor-management relations. In 1959, Mr. Huddart arrived in Georgetown as Bookers' first Personnel Director. His primary goal was to prevent the development of the kind of unified political-trade union complex the PPP sought to create. In its place he encouraged the growth of unions which strictly limited their activities to industrial concerns. With this purpose at heart, he

provided training and advice to Bookers' managers and many
trade union leaders.   Mr. Huddart considers his stay in
Georgetown a success.   When he arrived in the city in
1959, less than 10% of its labor force was unionized.
When he left in 1965, over 90% was.   Although explicitly
geared to discourage workers from using political methods
to defend their economic interests, Bookers' new
industrial-relations program did result in improved wages,
benefits and other conditions of employment.[18]   Employees of Bookers' Stores and urban industries enjoyed higher
wages, greater security and benefits than did their
counterparts in other privately owned businesses.

Indians, however, represented only 22 percent of the
urban population and for the majority still in the
country-side, Bookers' image as the loathsome Sugar King
did not change.   As you will recall, ex-indentured workers
did not (like the Blacks) divorce themselves from the
land.   With the help of the colonial state and sugar
planters determined to keep a large, cheap labor force
nearby Indians established relatively successful farming
communities throughout the sugar belt.   Bookers' expansion
and modernization of the sugar industry therefore had a
very different impact on the Indian population.

The urban population was largely indifferent to the
transformation taking place in the sugar belt.   Moreover,
while Black sugar workers, concentrated in the factory,
could benefit from the apprenticeship training school at
Port Mourant, Indian workers, concentrated in the fields,
were pushed out by the advance of mechanization.   Bookers
recognized that its mechanization program was throwing
people out of work at a time when unemployment was already
at a critical level; but, Bookers argued it was the
only way to make the industry internationally competitive
and profitable and to improve the standard of living of an
admittedly reduced labor force.   Between 1952 and 1960,
the number of sugar workers dropped to 20,480 from 28,352.
Thus, in less than a decade, the sugar industry, which had
traditionally supported the Indian community contracted
its workforce by 28 percent.[19]

Moreover, Bookers disclaimed any responsibility for
the relocation and welfare of its redundant labor force.
In Bookers opinion, "The problem of redistribution of the
population and the development of opportunity for other
work for persons who are redundant on the sugar estate is
one for Government--it is clearly beyond both the resources and the authority of the sugar industry."[20]   The
Indian community was outraged.   For generations, the
estate managers and overseers had controlled every detail
of the worker's life.   However miserable and inadequate,

the company had provided housing, health care and, when
necessary, food.   Change was occurring so quickly that
neither side had time to adequately adjust.   Jayarwardena
studied patterns of conflict on two of Bookers' estates in
the late 1950s. He observed the following:

> While, on the one hand, the laborers are as-
> serting their position in a nationwide class of
> "poor people" they have not relinquished their
> claims to protection by management from the
> vicissitudes of the labor market.   Again, while
> managers have renounced their former obligations
> to look after their laborers, they have not
> relinquished their claims to authority in extra-
> industrial affairs.[21]

Searwear offered a similar observation to the Georgetown
Chamber of Commerce. He told the group, largely composed
of Bookers executives, that despite the important changes
on the estates, the symbols of colonial exploitation, for
instance, the managers' big house and the separate living
and recreational facilities for senior staff, remained
highly visible.
    Furthermore, Searwear explained that the desperate
"land hunger" in the rural communities made it inevitable
that Indians would object to Bookers' post-war expansion.
Indian and Black farmers had always resented Bookers for
having legal title to the country's most fertile land much
of which the company did not even cultivate.   In this
regard, it is noteworthy that Bookers negative image also
persisted in the rural Black communities.   Watching
Bookers bring thousands of new acres under cultivation
stoked the fires of their long-term resentment.   In
addition, Bookers' new cane farming project, started in
1956, failed to satisfy the Indian's land hunger.   The
overwhelming negative features of the cane farming
scheme--the infertility of the scattered lands offered to
cane farmers at high rents, the exorbitant rates charged
for inputs, mechanical services and credits and the tight
paternalistic supervision of production--simply reinforced the hatred Indians felt for Bookers.[22] Likewise,
Bookers' description of its community development projects
in the new housing areas made for good public relations in
Georgetown but failed to impress the Indians living in the
communities who had to struggle with the gross inadequacy
of the company-provided community services and facilities.
    Nor was Bookers' image signficantly improved with the
Indian middle class.   The traditional Indian dream of
upward mobility was to become a large landowner, rice

miller or shopkeeper. Land and productive property (versus the Black dream of education) were held to be the keys to success. Thus, in contrast to members of the Black intelligentsia who leaped at the opportunities to join Bookers' top management, the Indian middle class, coming from a background of petty bourgeois ownership, held back. In 1959, five years after the start of the cadetship program, Bookers reported with obvious disappointment that, "Good progress with Guyanization is fairly obvious with Negro-Guyanese, but we have a long-term difficulty in bringing Indian-Guyanese into responsibility."[23] Eventually, Bookers did attract young, capable Indians who became devoted members of its senior staff, but, overall relations with the Indian middle class remained strained. Indeed, Indian merchants and shopkeepers became Bookers' most fierce competitors. They resented Bookers' expansion in the distributive sector and the rapid growth of Bookers' light industries. In response, they mobilized "race and political feelings in salesmanship"[24] and organized a separate Chamber of Commerce. This Junior Chamber of Commerce sought to reorient the colony's trade away from the United Kingdom to noncommonwealth territories. Furthermore, despite its capitalist ideology, the Junior Chamber of Commerce supported the PPP.[25] In its struggle with foreign capital, the Indian petty bourgeoisie formed an alliance with the anti-capitalist PPP.

Bookers' Suppression of Democratic Trade Unionism

Without a doubt, however, Bookers' most flagrant affront to the economic interest and democratic rights of the Indian population came in the form of the company's trade union policy. Bookers was determined at almost any cost to deny GIWU and its successor, GAWU, the Guyana Agricultural Worker's Union, its rightful status as the official bargaining agent for field and factory workers. GAWU was unacceptable because of its close affiliation with the "communist" PPP. Despite Jagan's intermittent conciliatory statements, both the PPP and Bookers understood that the party would eventually nationalize the industry and, in the meantime, could be depended upon to support demands for enormous wage and benefit increases. As an alternative to GAWU, Bookers put its power and resources behind the emasculated MPCA. According to Reno, this support was more instrumental to the union's survival than that of the American CIA and AFL-CIO or the British Trade Union Congress.[26] Nonetheless, Bookers claimed neutrality in the jurisdictional battle between the MPCA and the GAWU.

> Bookers' position in this dispute, as in 1963, is intensely difficult. It is our policy to work with established trade unions and to fulfill recognition agreements with them. And we recognize the right of workers to be represented by the unions of their choice: the problem is how fairly to elicit their choice. When there is conflict between union and union, and between Government and recognized unions, our actions fall between the crossfire of mutually exclusive forces.[27]

This was, of course, nonsense and today Bookers' executives admit that the company knew all along that the PPP's union had the overwhelming support of the workers. But, until the early 1970s, they were also convinced that GAWU's recognition would destroy their control of the industry.[28]

Therefore, an essential but unpublicized aspect of Bookers post-1953 program was an effort to destroy the PPP's union in the industry. As part of this effort, Bookers pumped new life into a union it had once bitterly opposed. The level of Bookers' support for the MPCA was in direct proportion to the perceived threat of Jagan and the PPP. For instance, when constitutional government was restored in 1957 and the PPP won re-election, Bookers decided to grant the MPCA the privilege of an automatic dues check-off. The BSE Board explained its policy reversal as follows, "If the financial underpinning, which it was expected check-off would give the MPCA, was not forthcoming, it seemed likely that Ishmael [MPCA's Secretary] would be forced to try and out-Jagan Jagan . . . . On the other hand, if he could have the strength which would spring from check-off, we could expect . . . his willingness to work constructively . . . would grow in proportion "[29]

In short, Bookers locked sugar workers into the membership of the MPCA. "As matters stood . . . MPCA representation could be changed only if a majority of workers went individually into company offices and signed withdrawals from the MPCA dues check-off. Workers were reluctant to do this since firings of PPP supporters were fresh in their minds, and mechanization made the fear of lay-offs even more real.[30]

In summary, Bookers' feverish activities to change its modus operandi and historic image were designed to defuse the political crisis in the aftermath of the constitutional supervision and remove the company from the center stage of the struggle for independence. However, because of the differential impact of the policies of Guyanization, modernization and company sponosored trade unionism on Black and East Indian communities, these goals were not realized. Blacks, most having long ago severed their ties to the sugar industry, were ready to accept the expatriate company at face value as a partner in the process of decolonization and economic development. The Indian population, still largely dependent upon the land and wages from the industry, remained bound in irreconcilable class conflict with Bookers. Thus it was inevitable that the most decisive battles in the struggle for control of the post-colonial state were literally fought out on the sugar estates and/or erupted from conflicts involving Bookers and the PPP's union in the industry.

Imperialist Intervention and Petty Bourgeois Factionalization

The PPP's unexpected victory raised the hopes of those who had fought and compromised to organize a unified movement to oppose colonialism. "For a brief moment sentiments of national unity transcending race or class or religious affiliation predominated. It was a time of excitement and hope containing the promise of a release of new energy and purpose for the building of a better future."[31] Unfortunately, this optimism and sense of revolutionary potential was quickly dissipated by the imposition of martial law. The Colonial Governor announced Emergency Orders which suspended all basis civil rights. Political and trade union leaders were either imprisoned or restricted by house arrest. With armed British troops patrolling the deserted streets "the old atmosphere of hopeless depression had returned, and the basic cleavages of race and class had begun to reassert themselves."[32]

The moderate leadership of the PPP increasingly denounced Jagan and his radical band and argued that the hard-liners had precipitated the crisis the "more reasonable" members had warned against all along. Thus, according to the moderates, the marxists had brought this indignity upon the nation because of their impetuous actions and needlessly provocative rhetoric. The British, aware of this growing internal division, moved skillfully

to ensure that it would erupt into a full-scale break.
But to achieve this goal the British needed time. Under
the recently suspended constitution the PPP was elector-
ally invincible. It was for this reason that the British
created a crisis where one had not previously existed and
imposed an interim government to rule until some acceptable solution to the marxist threat could be worked out.
In fact, three years after the arrival of British gunboats, Jock Campbell warned Bookers' shareholders the "I
have little doubt that were elections to be held now...the
Communist-dominated Peoples' Progressive Party would again
get a big majority...Nothing but a progressive, radical
and nationalistic party--with virile leadership--can
capture the imagination of the people."[33]

After suspending the Constitution in 1953, the
British advised that no further progress towards independence could be expected until a moderate political leadership could be found to replace the radicals epitomized by
Cheddi Jagan. Forbes Burnham was selected as their man
and the worst fears of radicals in the party were confirmed. He was described as a sensible socialist while
other PPP leaders were called communist extremists.
According to the report of the official commission of
inquiry into the suspension of the Constitution in 1953,

> ...we had no doubt that the socialists in the PPP
> were essentially democrats and that left to
> themselves their preference at all times would
> have been that the Party would pursue its con-
> stitutional objectives by straight-forward and
> peaceful means. We doubt however if they had the
> wit to see the essential difference between
> themselves and their communist colleagues or the
> ability to avoid being outmanoeuvred by them.[34]

Burnham did not miss the message concealed between the
lines. "An ambitious, devious man, Burnham played the
role of a modest and loyal British subject so long as
Whitehall remained in charge."[35]

In 1955, a moderate splinter group broke away from
original PPP in response to imperialism's siren call.
Along with Dr. Lachmansingh, the man who had previously
helped Dr. Jagan form the GIWU, Burnham split the tenuous
unity of the PPP's leadership and subsequently divided the
party's mass-based movement along radical lines. At first
there was considerably confusing among the PPP's grassroots supporters as Burnham and Jagan moved further and
further apart in the wake of the constitutional crisis.
Was the split really over tactics and ideology as the now

competing leadership claimed or was it racial? The
indications were very confusing since initially many of
the party's most active Black leaders stayed with Jagan
while Dr. Lachmansingh, a prominent Indian communal
leader, sided with Burnham. Even in a developed capitalist country wherein the working class has had the benefit
of a long tradition of class organization and political
participation it would have been difficult for the majority of the population to figure out what was actually
going on. In Guyana, wherein political mobilization was
new and racial segmentation was entrenched, it was
inevitable that confused Indian and Black workers would
define the problem in racial terms and react accordingly.
Since Burnham and Jagan both claimed to be socialists and
where nonetheless fighting each other the real problem,
they concluded, must be race. Racial suspicions and fear
of domination made more gut sense to the majority of
Guyanese than abstract distinctions between marxism-
leninism and social democracy.

The split in Guyana's independence movement and subsequent policization of racial divisions is an excellent
example of John Saul's theory of petty bourgeois politics
in the Third World.[36] Given the absence of a strong
national bourgeoisie or working class, the petty bourgeoisie emerges as the natural heir to state power after
the colonial authorities withdraw. Nonetheless, Saul
argues that as a class they are ill-prepared for the task.
The educated strata is rapidly expanding but it is a new
addition to the local class structure without deep roots
in the economy or established traditions. Furthermore,
the petty bourgeoisie is characterized by ethnic and
racial differences. It is therefore not surprising that
fragmentation readily occurs. The desire to possess the
state is overwhelming. In an underdeveloped society, the
state represents not only political power, it is the only
means available to develop an economic base for class
consolidation. As a result, political struggles channel
energies that would ordinarily flow into economic avenues
in a more materially advanced society. The struggle
for state power in Third World countries is therefore not
simply less institutionalized than in developed capitalist
societies, it is also more fierce.

In the ensuing competition for control of the post-
colonial state, the petty bourgeoisie fragments and the
rival factions then must appeal to whatever cultural
symbols or institutions will win them support. Thus, according to Saul, the principal cause of the politicization
of ethnicity and race in the Third World is the fierce
intraclass struggle for control of the postcolonial state

waged by members of the petty bourgeoisie -- not, as so
many have argued, the strength of primordial sentiments.

Once in open competition for control of the party
(Burnham's splinter group still called itself the PPP)
both Jagan and Burnham had to find a new basis for mass
mobilization. According to Despres, "Before the disintegration of the nationalist movement, mass support could
be mobilized primarily by appealing to the frustrations
that most Guyanese shared as a result of colonial domination. After disintegration, this approach continued to be
useful, but it was no longer sufficient....by 1958, the
East Indian and Afro-Guyanese cultural sections appeared
to represent the only bases of mass power accessible..."[37]
to the various factions struggling for control of the
post-colonial state. Burnham was the first to come to
this realization. In 1957, when elections were finally
held under a revised constitution which severely restricted the authority of the elected government, Burnham's
faction of the PPP was trounced. Jagan and his followers
won nine seats in the new legislature. Burnham's followers won only three. The remaining two seats went to
the National Labor Front and the United Democratic Party,
both right-wing organizations. Jagan was again called
upon to form the new government.

Burnham was dumbfounded. He reluctantly realized
that he would have to build an alternative political party
from the ground up. The PPP was clearly belonged to
Jagan. Peter Newman's analysis of the complex political
and racial dynamics which shaped the early development of
Burnham's new party, the People's National Congress (PNC)
is worthy of lengthy quotation:

> The defeat of the Burnham-led party resulted in
> its increased emphasis on African race-concious-
> ness. Haunted by the fact of a higher rate of
> Indian population growth...which will soon place
> Africans permanently into an electoral as well as
> population minority, it tried to create a Negro
> solidarity that would prevail at the polls
> against the existing very slight splinter over
> the appropriate brand of socialism. The PNC
> approach to the latter issue began its drift
> rightward as it merged for electoral and racial
> reasons with the moderate (and mainly African)
> United Democratic Party; but the rate of drift
> was checked by the ascension to the party from
> the Jagans' group of Sydney King, a Simon-pure
> Marxist and militant Negro leader, who became
> general secretary of the PNC. There was also an

> ultimate barrier to the extent to which the PNC
> could move right, formed by the political views
> of the Portugese and light-skinned middle class-
> es; any identification of the PNC with this group
> would at this period have meant the alienation of
> many poor, urban Negro voters, in addition to
> being personally objectionable to the Negro
> leaders of the party.[38]

Peter Newman concluded that the Black and lower class
character of the PNC's mass base would place an automatic
check on the rightward drift of the party and prevent the
leadership from forming an alliance with the Portugese and
the colored middle classes. He was only partially correct. The PNC continued to call itself a socialist party
but its pronouncements grew increasingly vague. Burnham's
address to the PNC's Annual Congress meeting on November
5th, 1961 is an excellent example of the nebulous quality
of the party's socialist position.

> It must be remembered that the PNC is a Socialist
> Party; by that I mean that the PNC draws its
> strength from the working people of this country,
> that the PNC is dedicated to the establishment of
> a system of social justice where the worker would
> get their just desserts, will get the rewards of
> their labor, and no longer will a few get the
> lion's share while the many get the jackal's
> pickings.[39]

Although never explicit, racial loyalty and solidarity became the principal vehicles by which the PNC mobilized popular support. The black population's fear of
domination by the numerically superior Indian population
was expertly manipulated. Their long held belief that
Guyana belonged to them and not to the Indians galvanized
poor and middle class Blacks into a political bloc. Race
conciousness indicated that the first line of defense had
to be held against the Indians. The colonial establishment was perceived as a lesser enemy. Indeed, the PNC
formed alliances with every major rightwing force in the
colony, i.e., the UDP, the CIA controlled trade union
movement and ultimately the Portugese and ultra-
conservative United Force.
By the time of the elections held in August 1961
racial solidarity had also become the keystone of the
PPP's mobilizing strategy.[40] The outcome of this election was seen as crucial since it was generally assumed
that the resulting government would soon lead the

country to independence. The Americans watched the process with particular concern. While Burnham and his activists concentrated their attention on the heavily Black urban areas, Jagan and his followers abandoned the cities and targeted their campaign on the rural areas. Jagan continued to denounce capitalism and advocate soviet socialism but his government adopted policies which were obviously designed to win the support of the Indian middle classes (i.e., teachers, civil servants, small business- men, large landowners) and small farmers. The most controversial policy was Black Bush Polder. The govern- ment spent millions of dollars to develop a new farming community which was turned over to land hunger peasants most of whom were, of course, Indian. The backbone of the PPP's elecoral support was still, however, the workers in the sugar industry. Unlike the small-scale farmers, sugar workers were concentrated in and around the large estates and thus easily organized. In fact, between the time of the demise of the GIWU in the mid-1950's (the union dissolved after Dr. Lachmansingh joined Burnham's maneuver to split the PPP) and the creation of the Guyana Agricul- tural Worker's Union in 1962, the PPP acted as an informal trade union for the workers.

Not surprisingly, Bookers was used as the quintes- sential example of capitalist exploitation. PPP spokes- persons reminded the sugar workers of what they already knew. Their sweat and poverty created the power and wealth of the foreign exploiter; therefore colonialism and capitalism had to be removed. The PPP was portrayed as the vanguard in this struggle. The PPP had always fought for sugar workers' right to democratic trade unionism, better wages and working conditions and ultimately the nationalization of the industry. On the other hand, they charged that PNC was in collusion with the Sugar King.

Bookers' Political Strategy

There is no hard evidence to support this charge. Instead the data indicates a number of points at which the interests and actions of Bookers and the PNC converged. Despite the rumors of collusion in the Indian community, Campbell and the new expatriates actually preferred Jagan to Burnham. Burnham was seen as an arrogant, brooding and race conscious man, opportunistically seeking the fruits of political power. He was an orator without equal but, Bookers' new managers detected an absence of any consist- ent political philosophy. In short, they felt he lacked integrity. On the other hand, the new expatriates

considered Jagan an affable and honest man with deep
political convictions.   Indeed, Campbell and his young
associates sympathized with many of the PPP's goals.
Nevertheless, they also found Jagan the embodiment of
incompetence, recklessness and shallow Marxism. According
to Tasker, Jagan would have wrecked the economy if his
party had continued in office.[41] Obviously, there were
also serious policy differences between the company and
the party.   Firstly, the PPP was determined to build its
political-trade union springboard in the sugar industry.
This meant that the party would do everything within its
power to secure recognition for the GAWU.   Secondly, the
PPP opposed membership for Guyana in the West Indies
Federation.   In its place, the PPP Government had
strengthened ties with Latin America, particularly
Cuba.   At the time Searwear observed, "There is a good
deal of evidence that such links are being cultivated as a
deliberate counter-weight to the attraction to the West
Indies Federation.   If this trend continues and there is
every indication that it will, it might subject trade
relations with the West Indies and the Commonwealth to a
degree of strain."[42]   Needless to say, this was totally
unacceptable to a United Kingdom based company with
operations scattered throughout the Commonwealth and
hoping to find lucrative business opportunities in the
West Indies Federation.   Thirdly, and most significantly,
the PPP was committed to a model of economic development
based on public ownership of large scale industry, in
part, their sugar industry.

Officially, Bookers took a position of political
neutrality during the turbulant years of the early 1960s.
Bookers neither publicly supported any political party nor
the attacks upon the PPP Government launched by the
opposition parties, the Trade Union Council or the Georgetown Chamber of Commerce.   But, of course, the major
political parties--the ultra-conservative, pro-capitalist
United Force, the ideologically ambiguous Peoples' National Congress, and the communist Peoples' Progressive
Party--were not all equally acceptable to Bookers.
Despite Bookers' official neutrality, deliberate efforts
to dissociate itself from the United Force and the London
Office's insistance that the company stand clear of the
subversive activities aimed at the PPP Government, political favoritism was apparent.   Old expatriates, local
whites, Portugese and Colored managers openly expressed
their support for the United Force.   More important from
the perspective of Bookers' future in Guyana was the fact
that the majority of the new Black senior staff members

were also members of the PNC's vanguard. For instance,
Harold Davis and George King were PNC activists. At the
time, Mr. Davis was a Personnel Officer for Bookers. He
rose to become the Personnel Director for Bookers' Sugar
Estates, Chairman of the Association of Guyanese Indus-
tries and Second Vice President of the Employers' Con-
federation. Today, Mr. Davis is the Chairman of the
state-owned sugar corporation, GUYSUCO. George King was a
manager of one of Bookers' Stores in the early sixties.
He was later appointed to the Board of Directors of
Bookers' Stores and served as an active member of the
Georgetown Chamber of Commerce. In 1973, Mr. King left
Bookers with the blessing of the company, to become the
Minister of Trade and Consumer Protection for the PNC
Government. In an ironic twist of history, the former
Booker Board member ended up on the Government's negoti-
ating team during the acquisition proceedings. Ms. Winnie
Gaskins was the editor of Bookers' News and chairperson of
the PNC. When the PNC rose to power in 1964, she became
Minister of Education. Finally there is Ptolemy Reid,
always second in command to Burnham, but rumored to be the
real power within the PNC regime. He was also once a
Booker Board member.

The picture was quite different from the perspective
of the Indian population. Any senior staff member who
supported the PPP was obliged to keep this information
private or face the loss of employment.[43] Obviously,
Bookers saw the PPP as inimical to its interests in the
same way that the Indian community saw Bookers. Thus when
Jagan sought to discredit the PNC with Indian voters he
alluded to the fact that many of the party's leading
personalities were members of Bookers' top management.
Just as importantly, this difference between the parties
indicated that the PNC was ready to be seen as allied with
Bookers:

> As a highly successful politician, Dr. Jagan must
> know whether or not such statements have force
> with his supporters. But, there is a second and
> even more important conclusion that I wish to
> draw. The fact that the Editor can hold office
> in a political party and that the repeated ac-
> cusations of Dr. Jagan does not appear to worry
> that party, must also mean that for another large
> section of the community, the bad historic image
> of Bookers has changed and given place to a far
> more favorable image, an image which can be
> reconciled to one of the main streams of our
> political life and aspirations.[44]

By the time of the elections in August 1961 the
lines of political conflict were sharply drawn. Jagan had
achieved some success in moderating the radical image of
the PPP by expelling from its leadership circle those he
claimed were guilty of ultraleft deviation and hence the
debacle of 1953. Indeed, by 1962, R.T. Smith observed
that "Very few of the people referred to as 'communists'
in the 1953 upset are still in the party; as a matter of
fact most of them are in highly respectable middle-class
occupations and have withdrawn from active politics."[45]
With marxism-leninism pushed to the background, the PPP
mobilized all of the various classses of the Indian
population into a voting bloc. Nevertheless, the party's
greatest strength was still in the rural districts,
particularly among the poor sugar workers. The foundation
of the party's platform was its opposition to colonialism
and capitalism best symbolized by the PPP's bitter and
unfilching opposition to King Sugar. Throughout the
campaign, the PPP leaders never missed an opportunity to
denounce the expatriate owners of the sugar industry and
to portray the PNC as their lackey.

> Consequently, a tide of Indian opposition to the
> PNC swelled across the estates. Indian sugar
> workers were convinced that a PNC victory at the
> polls would entail not only "black domination"
> but their continued exploitation. The opposition
> to Burnham was so strong among sugar workers that
> shortly before the 1961 elections it was almost
> impossible for the PNC to conduct a political
> meeting in the vicinity of the plantations."[46]

Meanwhile, the PNC had consolidated the Black population into a dependable electoral base. Burnham was
most popular among poor urban Blacks but the 1959 merger
with the conservative United Democratic Party also indicated that many middle class Blacks and Coloreds would
cast their ballot for the PNC. Moreover, the fact that
several influential Blacks had left Jagan's party and
thrown their support to Burnham meant that the PNC could
for the first time depend upon a substantial following in
the previously hostile Black rural communities.
The results of the elections were awaited with great
anxiety by all. The constitution, revised in 1960, gave
the locally elected government the power of internal
self-rule and it was assumed that this government would be
granted independence. The Americans watched Guyana's
progress towards independence with growing alarm. In the
wake of the successful Cuban Revolution and Castro's

defiant adoption of the soviet socialist model, the
Kennedy Administration decided that the virus of communism
had been set loose in the Western Hemisphere. Guyana was
its next target and the PPP embodied the disease. When
the votes were tallied the Americans realized that more
active involvement and outright subversion would be
necessary to prevent communist control of an independent
Guyana. The PPP had once again won control of the legislature with 20 seats compared to the 11 taken by the PNC
and the 4 going to the United Force (UF). With a clear
majority, the PPP formed the new Government.

It was obvious that the pro-capitalist UF had no
political future in the colony. The party of the upper
classes and Portugese had won only 16% of the vote. By
comparison, the PNC had taken 41% of the vote--only 2
percent less than the PPP. The PNC's party building and
electoral strategy had obviously been a success. The
remaining obstacle was the first-past-the-post electoral
system which gave an electoral seat to the party of the
candidate with the majority of votes in a district regardless of how slender his or her margin of victory was.
Smaller parties had denounced the system all along. Now
Burnham and D'Aguir, the leader of the UF, stepped up
their complaints. Together their parties had won more of
the popular vote than the PPP. They therefore claimed
that the PPP Government was illegitimate and that the
constitution should be changed to allow for a system of
proportional representation. Although still political
rivals, the PNC and the UF demanded this constitutional
amendment and new elections before independence be granted. Without external support, it is doubtful that the
opposition would have won; but, in this demand they had
powerful support from a formidable ally--the United States
government.

Despite his vaguely socialist rhetoric, Burnham and
the PNC were chosen by the Americans as the main political
means to stop the advance of Communism in Guyana. The
Kennedy Administration exerted consistent pressure on the
British to adopt proportional representation and to delay
the date of independence until such time as PPP could be
defected. While supporting Burnham, the key to the US
plan to subvert the PPP Government was its growing influence within the Guyanese TUM.

Unlike the British, the Americans did not seek to
divorce trade union activity from politics in Guyana.
On the contrary, scholarships to study at the American
Institute for Free Labor Development were awarded to
trade union leaders and activists on strictly political
grounds. Only those demonstrably hostile to the PPP

Government were given the privilege to visit and study in America. The Institute was formed in 1960 with the support of American business, trade union and government leaders. George Meany, the fanatically anti-communist leader of the AFL-CIO, became the Institute's first president and announced that the new arm of the American Labor movement would supplement the government's activities when formal diplomatic procedures made official action difficult. In short order, it became obvious that the AIFLD was frequently being used as a front for CIA subversive activities throughout the hemisphere. A variety of sources confirm the fact that the CIA actualized its plan to overthrow the Jagan Government through the good office of the AIFLD, the International Trade Sectionalists (ITS), and the ORIT affiliated unions in Guyana.[47]

During the period of heaviest US penetration of the TUM, Guyana received the highest proportion of labor scholarships in Latin America "...both on the basis of union membership and population of the colony."[48] According to Chase, each trainee was groomed to play a particular part in the plan "...to harass the Government by go slow, strikes, sabotage and other subversive activities, and if possible to overthrow the Government."[49] For their efforts, each trainee received $250 per month after returning to their posts within the TUM. Richard Ishmael, head of MPCA and reconstituted TUC, was responsible for coordinating their activities. Indeed, the AIFLD paid the salaries of the MPCA's full-time staff while the Public Services International covered the salaries of some of leading service union leaders in Guyana. Needless to say, the sweetheart sugar union and the public service unions formed the backbone of the anti-Jagan campaign. In addition to training abroad and heavy funding, American labor leaders were directly on-hand to ensure the success of their anti-communist policy. In the 18 months following the 1961 elections more US trade unionists visited Guyana than in the preceding 18 years.[50] In line with the subversive character of their mission they "...conducted courses and seminars...[on] how to fight communism and ways and means of opposing the Government."[51] Thus for the first time in this history of the Guyanese TUM, strikes came to be used for strictly political purposes. And, as Chase notes with bitter irony, we have to thank the Americans for this.

With the date of independence drawing near, the Americans called in their investment in the Guyanese TUM. In 1962 the TUC, now firmly under the control of the Americans, led a crippling strike against the Government.

The ostensible object of the strike was the new budget which Jagan submitted in late January. The real objective was to create such social disruption as to halt the movement toward independence. Faced with a severe short- age of development capital and Kennedy's rejection of their request for economic aid, the PPP Government sought to raise the necessary funds though a mandatory saving program for workers, increased tax rates on the private sector, particularly foreign capital, and increased taxes on luxury consumer items. The New York and London Times called the budget a fair and bold effort to generate funds for economic development. Bookers, the largest expatriate business in the country, felt that "It clearly was, in intention, a serious attempt by the Government to get to grips with the formidable economic problems of the country by a hard progam of self-help. It was a radical budget (what have the people of British Guiana got to be con- servative about?) but not confiscatory..."[52]

Most of the business community did not accept Jock Campbell's liberal assessment. The then privately owned Daily Chronicle announced in bold headlines on February 2, 1962 that the "Budget is Marxist. A vindictive and mali- cious spirit prowls throught the budget." In actual fact, the budget had been conceived by Nicholas Kaldor, a Cam- bridge University economist and tax expert. Not to be deterred by such incidentals, the anti-Jagan unions claimed that the budget was anti-working class and com- munistic! Members of the Georgetown Chamber of Commerce supported the TUC's strike call by locking the doors to their businesses and paying striking workers. According to Ronald Rodosh, the CIA was on hand providing advice on how to organize the strike and providing funds, food and medical supplies to sustain its momentum. Pro-Burnham dike and public employees unions were especially favored recipients of such assistance.[53] The opposition parties led by Burnham and D'Aguir naturally lent their support to the movement to overthrow Jagan. Carrying placards which read "Choke and Rob Budget," "Slavery if Jagan gets Independence Now" and "Down with the Government" they led mass rallies around the public buildings. Moreover, when the predicted violence and arson broke out, both Burnham and D'Aguir ignored the Governor's request that they seek to calm their supporters. In a similar vein, the unions continued their opposition even after the Government agreed to remove from the budget those provisions they had originally objected to.

British troops finally had to be brought in to quiet the rioting and restore order. The PPP was forced to withdraw the budget. More importantly, from the perspective of the opposition and the Americans, they had demonstrated that Jagan could not, despite his constitutional authority, effectively govern the country.

With its grip on state power quickly slipping, the PPP sought to strengthen its hand in the labor movement. The original plan to capture control of the TUM was clearly no longer practical. The Americans had preempted that move. Nevertheless, the party's leadership still thought it was possible to win recognition for its union in the sugar industry, the GAWU. If the party somehow managed to stay in power, having gained recognition for the largest trade union in the country would greatly increase the PPP's control of economic policy and enhance the party's leverage within imperialist dominated TUM. If, on the other hand, pro-imperialist forces succeeded in overthrowing Jagan's Government, the PPP would be much more powerful as the opposition party. A recognized GAWU would grow in strength and be able to hamstring any Government the imperialist might impose. With these considerations in mind, the GAWU issued a strike call in early 1963 after the Sugar Producers' Association rejected its request for recognition.

Bookers, in lieu with its long-standing policy, claimed neutrality but nonetheless managed to support every proposal made by the MCPA. Most importantly, while proclaiming their respect for the right of workers to be represented by the unions of their choice, Bookers staunchly opposed the new Labor Relations Bill submitted by the Government in late March as a means to resolve the dispute. The Labor Relations Bill of 1963 was practically identical to the 1953 bill which triggered the suspension of the Constitution. In brief, it sought to guarantee workers' rights to select their union representative by secret ballot and mandated recognition of the union winning such a poll.

With the LRB once again before the parliament, GAWU called off its strike and awaited legislative action. In 1953 the TUC supported the LRB. Now, firmly under the direction of the CIA, the TUC leadership denounced the bill as a proposal to "Castorize" the unions and destroy democratic trade unionism in the colony. Forbes Burnham, who in 1953 eloquently presented the LRB to parliament, now charged that the bill was evidence that Jagan was trying to turn Guyana into a "Soviet Satellite."[54]

In a bold faced political move, the TUC called for a general strike to protest the proposed LRB. The TUC leadership claimed that the draft legislation would, by calling for the creation of a Government appointed body to oversee union elections, ensure that PPP leaders

controlled election results. When, however, the PPP
proposed that the International Labor Organization be
brought in to supervise the poll in the sugar industry,
the TUC leadership still refused to drop its opposition.
Obviously, as in 1962, the Government's immediate action
merely served as a pretext to disrupt the economy and
hopefully bring down the democratically elected socialist
Government.

On April 17, two Americans flew into Guyana and held
all night meetings with Ricard Ishmael, the head of the
MPCA and TUC. The next day, Ishmael called for the
general strike which was to drag on for an incredible 80
days. Reflecting his renegade status among sugar workers,
only one tenth of the workforce in the industry responded.
This however did not break the momentum of the strike.
Bookers and Demerara, the two largest sugar producers,
locked their workers out. Despite the heavy financial
losses involved, Bookers resolved that defeating the GAWU
was more important than immediate profits. The public
service unions comprised mostly of Black, pro-Burnham
workers enthusiastically supported the strike as a means
to oppose the political power of the Indian population.
Mr. Romualdi, director of the AIFLD, assigned six graduates of his program to the strike coordinating committee.
Most importantly, the CIA via its connections with the PSI
pumped in a constant supply of money and materials to ease
the burden on the striking workers--initially, between
$30,000 and $50,000 a week; by the end, $130,000 weekly.
By CIA standards this was a meager amount but to many
Guyanese workers the subsidy represented more than their
usual wages. The US intervention was crucial. "This
money enabled the unions to erect new demands each time
that Jagan gave in on their original ones."[55] Helpless
before this alliance between its internal opposition and
the Americans, the PPP was forced to withdraw the labor
bill. The strike which began in April was finally called
off in July.

Just as the opposition had desired, the British used
the strike and resulting social and economic upheaval as
an excuse for postponing independence. Yet another
constitutional conference was held in October to discuss
the positions of the major contenders in the struggle for
control of the postcolonial state. Burnham and D'Aguir,
strengthened by the recent display of their disruptive
power, were uncompromising in their demand that the
electoral system be changed to proportional representation. Jagan, considerably weakened by his Government's
inability to maintain order, still insisted that the
existing electoral system found in virtually all British

colonies and former possessions not be tampered with. On the other hand, the PPP requested that the legal voting age be lowered to eighteen since the higher Indian birth rate translated into a younger population. Given the intransigence of each side, an impasse emerged. Finally, the three men agreed to submit the outstanding issues to the Colonial Secretary, Duncan Sandys, and to abide by his decisions. For Jagan this was a fatal mistake. While the Colonial Office had in its usually aloof and bureaucratic manner maintained a posture of impartiality, the British (under enormous pressure from the Americans) were looking for a way to get Jagan out of office. Jagan handed them the means on a silver platter at the October Conference. For some inexplicable reason, Jagan obviously still trusted in the purported fairness of the British.

After consultations with the Americans and Bookers, Duncan Sandys announced "his" decision. The electoral system would be changed to proportional representation. The voting age would not be lowered. And, as the last bit of insurance against a PPP victory, the date for independence would be set after new elections were held under proportional representation. After this constitutional coup, the PPP in desperation called for national demonstrations to protest Sandy's decision. This failing to reverse the decision, the PPP mobilized its most powerful resource, the GAWU.

Using the tactics of the opposition, the PPP provoked a strike in the sugar industry. In rage and indignation, 14,000 sugar workers risked their jobs and signed letters resigning from the MCPA and naming the GAWU as their bargaining agent. It was a decisive move. The MCPA with the support of the companies had claim that the names on their membership books representated the workers' choice. That fiction was now shattered. Still, Bookers refused to budge. A strike call went out on February 17, 1964. Then suddenly the politicized trade union struggle was transformed into a bloody racial war when the companies brought in Black scab labor. "In any other country with a homogeneous population" Jagan pointed out, "it would have been strikers battling against scabs in an industrial dispute. In Guyana, because the strikers were mainly Indian and the scabs mainly African, an industrial dispute turned into a racial war."[56] The conflagration in the sugar belt quickly engulfed the entire country as rumors spread and Indian and Black communities exchanged deadly attacks. According to Aston Chase's firsthand account, "Bombings, shootings and savage assaults were the order of the day. Arson was rampant . . . Physical partition in

certain areas came into being with the rapidity of lightning."[57]   The carnage lasted six months.   In the end,
176 people were dead, 920 were injured and 2,668 families
were forced to relocate.   Property damage was placed at
$4.3 million.[58]

This tragic episode represents a watershed in Guyana's social and political development. It finalized the
process of racial polarization which began with Burnham's
departure from the PPP and his creation of a separate
Black party. As fate would have it the process of polarization culminated in the sugar industry. Writing during
the midst of the crisis, Philip Reno observed that, "The
history of British Guyana has been the history of struggle
between the sugar plantocracy and Guyanese sugar workers,
and it is therefore fitting that British and American
control of Guyana comes to a showdown in a strike by sugar
workers."[59]   The two racial groups realized their worst
potential and buried their recent unity in attempted
genocide. Hereafter, racial hostility and fear became the
most important features of Guyana's political life. Not
surprisingly, Jagan held the PNC responsible for much of
the terrorism. He claims that, "the reign of terror in
Georgetown was halted when the police accidently raided
the hotel room of a PNC activist . . . and found arms,
ammunition and explosives."[60]   What Jagan does not
stress is the fact that the sugar strike began only after
major demonstrations led by the PPP had failed to force
the Colonial Secretary to rescind his decision to impose
proportional representation.   Clearly both parties were
exploiting the sugar workers' struggle for their own
political ends.

Elections were held under the new system of proportional representation on December 7, 1964. The PPP won 24
seats and the PNC and UF won 22 seats and 7 seats respectively. Despite Burnham's often repeated claim that he
would never enter an alliance with the UF, the PNC formed
a coalition Government with the ultra-conservative,
pro-capitalist party of the Portugese and light-skinned
middle classes.   Obviously, Burnham preferred to be a
senior partner in a neocolonial Government rather than a
junior partner in a popular Government headed by the
PPP.

The decision to form a coalition Government with the
representatives of local and foreign capital merely
cemented a rightwing alliance already well established.
As we have seen, Burnham and his party were more than
willing to be associated with the expatriate owners of the
sugar industry. Key PNC leaders were members of Bookers
local top management. Many of them moved directly from

the company into the new Government. We believe this transfer of personnel from the company to the Government helps to explain the strength and the longevity of the alliance between Bookers and the PNC which first emerged in the late 1950s. From the perspective of the PNC leadership, Bookers was not an alien capitalist firm ripping off the country. On the contrary, due to the company's program of Guyanization, they had become privileged members of the Bookers team and remained closely identified with the company.

The alliance also made sense from the perspective of the company. Bookers had long recognized that "Nothing but a progressive, radical and nationalistic party--with virile leadership--can capture the imagination of the people."[61] Secondly, the PNC, despite radical rhetoric was obviously pro-Western and open to pursuing a capitalist development strategy. The party was a strong supporter of the proposed West Indies Federation and continued membership for an independent Guyana in the Commonwealth. Whereas the PPP labored to develop ties with Latin America and the Eastern bloc, Burnham was content with the status quo. "We all know that developing countries are seeking economical aid and assistance, and if there is a nation with which there have been ties, I can see no objection to seeking such aid and assistance therefore."[62] Given the choice between two rival imperialist blocs, the PNC sided with the Western industrialized nations. Unlike the anti-capitalist PPP, the PNC promised to develop Guyana in partnership with multinational companies and local private capital.

In addition to this generally pro-capitalist orientation, the PNC and Bookers shared specific interests in the sugar industry. Both were equally committed to the continued suppression of GAWU. On the one hand, Bookers was convinced that GAWU's recognition would mean economic ruin and the company was prepared to forego immediate profits in order to ensure its survival in the industry. On the other hand, the PNC recognized that GAWU's victory would be a tremendous boon to the PPP. Thereafter, Jagan would attempt to gain control of the National Trade Union Council and with that type of trade union strength, the PPP would be virtually in control of the economy and politically invincible. Furthermore, as the ruling party, the PNC gained economic reasons for opposing the GAWU. After bauxite, sugar was the next largest contributor to the Government's treasury. Production stoppages not only threatened company profits, they also reduced tax revenues.

It is therefore not necessary to posit collusion, as the PPP has done, to explain the close relationship which developed between the PNC and Bookers. The PNC and Bookers had similar goals which were obvious to all. Seeking similar ends they pursued similar policies. Throughout the long struggle for independence, there is no evidence of back-room dealings between the two. What is more surprising is the open manner in which the Black leaders embraced the white, expatriate firm after their brief and disastrous alliance with the Indian population. This we have seen was possible because of the largely urban character of the Black population, its devotion to British culture and the success of Bookers' Guyanization program among the Black middle class. When you add to these social structural features the common fear of the PPP/GAWU, the partnership between the PNC and Bookers makes perfect sense. This close partnership between Bookers and the PNC would last for roughly twenty years.

From the beginning, the coalition between the Black and Indian leaders of the independence movement had been fragile. Past, sporadic struggles to organize the working class into an effective labor movement had not produced significant or lasting gains. Black and Indian workers were still largely segregated in different economic sectors and occupations. There was no tradition of working class solidarity. Indian and Black workers and peasants defined themselves first and foremost as members of competing racial groups in the country's color-class hierarchy. Their unity within the early PPP rested upon their loyalty to their respective charismatic leaders rather than a commitment to class struggle and socialism. Consequently, when Jagan and Burnham split there were no structural factors to prevent the degeneration of the socialist oriented movement into an internecine struggle for control of the post colonial state. The socialist ideology had been imposed from above and lacked roots in the consciousness or experience of the people.

Thus the unity of the Indian and Black population had been primarily dependent upon an agreement between their leaders drawn from the pettly bourgeois strata. When this agreement crumbled under the pressure applied by the imperialist powers (which we take as a given in any progressive movement) both Burnham and Jagan pushed their socialist principles into the background and mobilized support on the most readily available basis, racial loyalty. To do otherwise would have meant accepting electoral defeat. Paraphrasing Amilcar Cabral, we can conclude that neither Burnham nor Jagan was ready to commit political suicide for the sake of working class

unity.   Moreover, from the perspective of most Guyanese,
race was a much more real and vibrant energy than class.

NOTES
1. Bookers' Sugar, supplement to Booker McConnell, Ltd.,
    Annual Report and Accounts 1954, p. 65.
2. Brian Scott, "The Organization Network: A Strategy
    Perspective for Development" (Ph.D. dissertation,
    Harvard University, 1979), p. 122.
3. Quoted in Bookers' Sugar, p. 94.
4. Booker McConnell, Ltd., Annual Report and Accounts
    1960, Statement by the Chairman.
5. Booker McConnell, Ltd., Annual Report and Accounts
    1970, Statement by the Chairman.
6. Scott, pp. 166-167.
7. Quoted in Bookers' Sugar, p. 90.
8. Bookers' News (Georgetown), 17 December 1975.
9. Booker McConnell, Ltd., Annual Report and Accounts
    1966, Statement by the Chairman.
10. The Guyana Graphic, 30 August 1967.
11. Interview with Anthony Haynes, Chief Executive
    Officer Booker McConnell, Ltd., London, England,
    August 1980. Mr. Haynes' comment is surprisingly
    similar to the point raised by Clive Thomas in his
    biting criticism of the program: "The white expat-
    riate is no longer visible . . . . The workers direct
    their grievances to the local man, and this it is
    hoped would divert their attention from the expa-
    triate owner." Clive Thomas "Meaningful Participa-
    tion: The Fraud of It," in The Aftermath of Sover-
    eignty, West Indian Perspectives, eds. David Lowen-
    thal and Lambros Comitas (Garden City: Anchor Press,
    1973), p. 356.
12. The Guyana Graphic, 30 August 1967.
13. Ibid.
14. Quoted in Scott, p. 148.
15. Interview with Anthony Tasker, former Public Rela-
    tions Officer, conducted at Booker McConnell head-
    quarters, London, England, August 1980.
16. Lloyd Searwar, "Bookers Guiana--A Study in the
    Development of an Image and the Implications for
    Independence," n.d., n.p. (Mimeograph.)
17. Searwar, p. 7.
18. Interview with John Huddart, former Personnel Direct-
    or of Bookers' Non-Sugar Estates Businesses, conduct-
    ed at Booker McConnell's branch in Gerrards Cross,
    England, August 1980.
19. Scott, pp. 134-143.
20. Bookers' Sugar, p. 69.

21. Chandra Jayawardena, Conflict and Solidarity in a Guyanese Plantation (London: Athlone, 1963), pp. 26-27.
22. Scott, pp. 183-192.
23. Quoted in Scott, p. 147.
24. Swearwar, p. 10.
25. Ibid.
26. Philip Reno, "The Ordeal of British Guiana," Monthly Review 16 (July-August 1964); 93-94.
27. Booker McConnell, Ltd., Annual Report and Accounts 1963, Statement by the Chairman.
28. Scott, p. 173.
29. Quoted in Scott, p. 177.
30. Reno, p. 52.
31. Smith, p. 171.
32. Ibid., pp. 171-72.
33. Booker McConnell, Ltd., Annual Report and Accounts 1955, Statement by the Chairman.
34. Quoted in Cheddi Jagan, The West on Trial: The Fight for Guyana's Freedom, rev. 2d ed. (Berlin, German Democratic Republic: Seven Seas Books, 1975), p. 162.
35. Penny Lernoux, "Jonestown Nation," The Nation 15 November 1980.
36. John Saul, The State and Revolution in Eastern Africa. (New York: Monthly Review Press, 1979).
37. Despres, p. 221.
38. Peter Newman, "Racial Tensions in British Guiana," Race 3 (May 1962): 31-45.
39. Quoted in The Chronicle (Georgetown) 5 October 1980.
40. Despres, pp. 222-250.
41. Interview with Mr. Tasker, ibid.
42. Searwar, p. 8.
43. Ibid.
44. Searwar, p. 2.
45. Smith, p. 182.
46. Despres, p. 244.
47. For instance, see Radosh, American Labor and U.S. Foreign Policy. Reno "The Ordeal of British Guyana"; Jagan, The West on Trial.
48. Chase, p. 294.
49. Ibid.
50. Chase, p. 272.
51. Ibid., p. 291.

52. Booker McConnell, Ltd., Annual Report and Accounts 1961, Statement by the Chairman.
53. Radosh, p. 404.
     1963, Statement by the Chairman.
54. Ibid.
55. Ibid., p. 403.
56. Cheddi Jagan, The West on Trial: The Fight for Guyana's Freedom, rev. 2d ed. (Berlin, German Democratic Republic: Seven Seas Books, 1975), p. 306.
57. Chase, p. 305.
58. Jagan, p. 311.
59. Reno, p. 65.
60. Ibid.
61. Booker McConnell, Ltd., Annual Report and Accounts 1955, Statement by the Chairman.
62. Parliament House, Debates, 21 June 1966, p. 221.

CHAPTER 4

The Radicalization of the PNC's Development Strategy

        The purpose of this chapter is to examine the trans-
formation of the development strategy pursued by the
postcolonial state in Guyana. The focus of the analysis
will be on (1) the nature of the shifts in the state's
changing developmental policy, (2) the related changes in
governing alliance including the structure of the ruling
party and (3) the role of external factors in shaping the
state's development strategy. During the decade of the
1970s the development strategy of the Guyanese state
veered sharply to the left. Starting from a neo-colonial
orientation, the Government adopted an indigenous brand of
socialism, nationalized eighty per cent of the economy,
embraced Marxism-Leninism, and declared the paramountcy of
the ruling party. Our goal is to understand the domestic
and international determinants of this radical policy
shift and to interpret its implications for the goals of
socialism and democracy in Guyana.

The Formation of Guyana's Neo-Colonial Alliance

        In 1966 Guyana was finally granted independence from
England. The state was entrusted to the coalition Government formed by the Peoples' National Congress and the
United Force, headed by Forbes Burnham. The PNC still said
it was a socialist party but the leadership carefully
avoided any clear statement concerning how it intended to
move the country away from exploitative class relations
and its peripheral position within the world capitalist
system. The UF, on the contrary, was openly committed to
a capitalist development strategy based on a close relationship with the former motherland and the United States.
Peter D'Aguir, the leader of the UF, firmly believed that
"It would be unrealistic for the government to involve
itself into excursions in private industry. That should
be left to those who are experts. British Guiana has got
to attract capital, but this will be impossible if the
government involves itself in commercial affairs."[1] Mr.
D'Aguir became the Minister of Finance in the independence
Government and supervised the design and implementation of
a classical, neo-colonial development strategy. Sir
Arthur Lewis, the champion of the Puerto Rican Model in
the British Caribbean, became the chief architect of the
first, post-colonial development plan.
        Economic policy reflected Minister D'Aguir's reliance
upon private foreign and local capital to be the engine
of economic development. The expatriate owned bauxite and

sugar industries were given handsome incentives to increase their investments and expand production. Capital
gains taxes were reduced, depreciation allowances were
greatly increased, prestige advertisement by businesses
was made tax deductible and most importantly, exchange
controls were lifted making for the free flow of capital
and profits. Final assurance was given to private investors in the Independence Constitution which guaranteed that
the government would pay prompt and adequate compensation
for any property acquired by the state.[2]

Dr. Lewis' Development Plan for 1966-72 reflected the
economic assumptions and strategies he had championed
since the early 1940s. In brief, Dr. Lewis insisted that
economic development in the West Indies was dependent upon
the large infusion of foreign capital to be invested in
light, labor intensive manufacturing operations producing
for the export market. The domestic savings of Caribbean
countries are insufficient to fuel this type of industrialization and the governments are not able to borrow
all of the necessary amount. Thus, according to Lewis,
pragmatic politicians would have to be willing to accept
responsibility for "wooing and fawning" potential foreign
investors. Lewis supported this argument by pointing to
the following unpleasant considerations. First, few
manufacturers are eager or even interested in investing in
the Caribbean. Second, not even the low wages of Caribbean workers are sufficient to attract them. Therefore,
governments must be ready to offer temporary monopolies,
subsidies, tax holidays, tariffs protection, and other
reasonable inducements. Dr. Lewis realized that such a
policy would be unpalatable to many nationalist leaders
but he warned that refusal to face the facts would "do a
grave disservice" to the people of the region.[3] The
PNC-UF Government was not difficult to persuade.

The 1966-72 Development Plan rested on the premises
of Dr. Lewis' philosophy of Caribbean development. The
G$294 million Plan (only G$50 million of which was to
be internally generated) directed the bulk of public
investment into infrastructural projects. Areas of
productive investment were carefully left to private
capital. The magic of the free market was trusted to
beneficiently guide the development process for the
general welfare of society. In regard to the role of the
private sector Kempe Hope noted that the Lewis Plan ". . .
did not in any way project or quantify the expected
contribution . . . . Private enterprise was expected to
willingly help so that the economy would flourish and
expand. No attempt was made to show what kind of help was
required or desired from private enterprise in order that

the goals of the Plan be realized." As to infrastructural
projects, " . . . no consideration was given to the
expected returns of such projects over the long run."[4]
Not even the Tudor monarchs had had such faith in unguided
capitalist development.

Maintenance of the Colonial Framework of Labor Relations

Whereas the PPP fought to destroy the colonial
framework of industrial relations and thereby gain the
control of a liberated trade union movement, the PNC
formed an alliance with those forces in the society
determined to defend the exploitative relationship existing between labor and capital. Therefore the early labor
policy of th post-colonial state bore a close resemblance
to the labor policies of the former colonial state. The
most significant indicator of this continuity of labor
policy was the decision to retain the British voluntarist
framework with its implicit anti-working class bias. The
Critchlow Labor College, responsible for the ideological
training of the working class, continued to receive its
financing and direction from the American Central Intelligence Agency.[5] Employers retained complete discretion
in regard to union recognition; no legally binding procedures for the resolution of inter-union jurisdictional
disputes were prescribed, and collective bargaining
agreements remained legally unenforceable. Thus after
decolonization labor still held only the double-edged
strike weapon in its highly unequal struggle against
capital. Nonetheless, workers continued to fight against
their exploitation. Between 1965 and 1970 there were more
than 900 strikes in Guyana and over   1,000,000 man-days
lost.[6]   From the perspective of labor, management and
the government, the situation was untenable. A small,
underdeveloped economy struggling to increase productivity, attract capital investments, diversify production and raise the standard of living simply could not
afford such a high level of labor disruption. In addition
to the profits and wages being lost, the Government was
losing desperately needed revenues and crucial foreign
exchange earnings. Something had to be done but the
question was, what?
From the time the PNC assumed office in 1964, it
diligently tried to prevent the strikes which helped
to frustrate its development plans. However, as the
head of Government the PNC was not willing to upset the
configuration of forces between management and labor in
such a way as to legitimize labor's wildcat and/or

militant leadership. For instance, minimum legislative
protection for unions such as exists in the U.S. would
correct those aspects of the British voluntarist tradition
which result in Guyana's high level of strike activity but
it would also undercut the interest of employers and the
renegade leadership of the TUM which has always supported
the PNC.[7] Most obviously, the PNC did not want to
create the legal framework which would lead to GAWU's
recognition. Under no conditions was the PNC ready to
have the PPP (via GAWU) gain control of the Trade Union
Council (TUC), the official mouthpiece of organized
labor in the country. Thus, in place of legislative protection for democratic union elections and mandatory
recognition, the Government sought to place a legal ban
on strikes and to empower its Minister of Labor with the
authority to impose compulsory arbitration. The Government repeatedly sought to enact its strike ban but even
the coopted leadership of the TUM refused to go along and
threatened serious protests. The PNC was stuck between
Scylla and Charibdis. On the one side, it faced the
persistence of crippling strike activity; on the other, it
faced the more dreaded possibility of the domination of
the TUM by its political opposition. Trapped, they did
nothing at all. The colonial framework of labor relations
remained intact.

The Failure of the Classical Neocolonial Model

The fate of any development model is determined by
its results. Hopes run high when a new government takes
office but the leaders have a short time in which to prove
themselves. In accordance with the Lewis Plan, the PNC-UF
Government spent millins of dollars developing the country's infrastructure and soliciting private investors.
Guyana was not as "lucky" as Puerto Rico where huge
investments of American capital were made. The promised
infusion of large-scale foreign investments did not take
place. The taint of communism clung to the body politic
in Guyana and foreign capitalists chose the hands of
docile and cheaper labor elsewhere in the Third World.
Nor did local private capital make good use of the Government's incentives to increase investments and diversify
the economy. The problems of unemployment, underemployment and inflation continued to grow worse. To further
aggravate the situation from the perspective of the
Government, national elections were scheduled for December
1968. The PNC feared that it would lose if free and fair
balloting was allowed to occur. Instead the party chose

to rig the elections,[8] drop its coalition partner and jettison the 1966-72 Development Plan three years before it was due to expire.

The Radicalization of the PNC's Development Strategy

Thus began the PNC's transformation into the type of regime Juan Linz has characterized as authoritarian.[9] Since the 1968 elections the ruling party has inexorably moved to limit the mobilization of opposition without however seeking to totally eliminate dissent. On the contrary, the PNC has quite creatively maintained institutions symbolic of genuine competitive politics while stripping them of their intended substance. Thus we find formally recognized political parties and a de facto one party state; constitutionally guaranteed elections and regular electoral rigging; an official revolutionary ideology and a pattern of rule reliant upon intimidation, apathy and mass mobilization manipulated from above. Linz emphasizes the fact that authoritarian regimes are generally committed to modernizing society and frequently profess some brand of revolutionary ideology. However, the ideas which guide these regimes are vague, emotional and non-systematic. Thus although the authoritarian regime seeks to legitimate its rule on the basis of efficiency and economic progress, they are generally non-dynamic. The narrow political base of these regimes forces them to depoliticize the masses. In our view this is the fatal contradiction of the authoritarian regimes. Economic development of underdeveloped societies demands enormous commitments of energy and enthusiasm from the labor force and an equal commitment to restrict immediate consumption. This type of personal and collective investment will only take place when the mass of the population is actively in control of the development process and/or following a very popular leadership. Unfortunately, our analysis reveals that the PNC regime lacks either of these characteristics.

In a grand gesture the PNC renounced its alliance with the local bourgeoisie. The economic failure of the PNC-UF Government was blamed on the obdurant and capitalist nature of the UF. According to Forbes Burnham, the PNC wanted to elevate the small man but the UF was determined to keep him down. The PNC wanted to break the ties with the colonial past and establish a republic, but the UF insisted on keeping the Queen of England as Head of State. Most importantly, the PNC wanted to start the journey along the socialist path of development and now freed from its distasteful but expedient partnership, the party would be faithful to its true socialist nature.[10]

The constellation of local class, racial and political factors facilitated the radicalization of the PNC's development strategy. When the PNC decided to abandon the traditional neo-colonial strategy, the local bourgeoisie which had formed an important part of that alliance was virtually powerless to resist. As a despised racial minority (they are mainly Portuguese), the national bourgeoisie can not effectively act without local allies. Once the Black political leadership turned to the left, the UF had no other local resource. Meanwhile, the East Indian majority remained loyal to the PPP, the communist party led by Cheddi Jagan.

The PPP represents Guyana's radical past and according to its leaders and the Americans the hopes of its communist future. Dr. Jagan has not been able to bridge the racial division between Indians and Blacks which characterizes Guyana's political structure but the PPP has kept a communist critique and alternative to PNC policy before the public eye. The party's newspaper, The Mirror is popular among Blacks as well as East Indians. More importantly, the PPP has seen it as a duty to support and encourage every leftist departure undertaken by the ruling party. In Guyana, it is the right and not the left which must justify its existent. The poverty striken Indian and Black workers applaud every attack on pro-capitalist forces which they identify with the Portuguese, Coloreds and imperialist powers. Thus, if the theory is correct, the internal structure of Guyanese society favors a non-capitalist development strategy. Needless to say, external factors represented by the world capitalist market do not; and, as we will subsequently see have effectively checked the leftist aspirations of the PNC.

On February 23, 1970, the PNC proclaimed Guyana the world's first Cooperative Republic. Cooperativism is the PNC's brand of socialism. According to the party, it is an indigenously derived means for transforming the colonial economic structure based on exploitation and private property into a social system based on equality, justice and collective property. To buttress this claim, party ideologues have interpreted every previous anti-slavery and anti-colonial movement as evidence of the natural cooperativism of the Guyanese people.[11] In addition, cooperativism is praised as a decentralized, voluntaristic and gradual approach to revolutionary change. The Soviet model advocated by the PPP opposition which is based on a high degree of centralization, bureaucracy and state planning of the economy is rejected. The PNC, which had

so recently rigged the elections, said it favored a
democratic socialist model! Citizens who had discovered
that their names were mysteriously removed from voter
rolls or that a proxy vote had been cast in their behalf
must have been incredulous as the Prime Minister declared
that " . . . the small man's economic power must be in
direct proportion to his political power. Otherwise
independence and republic status are meaningless . . . .
The co-operative has been chosen because structurally it
lends itself to organizing the small man in groups for
economic purposes. By definition also, it is democratic
in that the principle of one man one vote, instead of one
share one vote, ensures the value and importance of the
man as distinct from those of mere coin."[12] Despite the
obvious contradiction between electoral fraud and the
proclamation of a democratic socialist model, some
of Guyana's most respected intellectuals argued for
supportiveness and patience as the Government worked out
the particulars of the policy they hoped would be "bold
and imaginative." It was felt that the moderate socialist
strategy advocated by the PNC had more chance of success
than the radical approach of the PPP.[13] A twist of
irony was added to the situation when the pro-Soviet PPP
denounced the program as fraudulent since socialism could
not, it claimed, be built in a bureaucratized and increasingly repressive environment such as was rapidly
unfolding in Guyana.[14].

Clarification of the actual economic policies to be
pursued were slow in coming while the symbolic changes
came quickly and were emotionally satisfying to most
Guyanese. Pictures of the British Royal Family were
removed from the legislative chambers, honors and titles
associated with the colonial motherland were replaced by a
national system of merit recognition, judicial gowns and
wigs were disgarded and the tropically inspired "shirt-
jac" became the fashion. No similarly dramatic attempt
was made to explain how the private, public and cooperative sectors would work together to ensure the socialist
development of the economy. The dynamics of tri-sectoral
development, especially how the fledgling cooperative
sector would come to dominate the private and public
sectors remained a mystery.

The cooperative movement was not new to Guyana. The
colonial government had introduced cooperatives into the
colony during World War II to compensate for disruptions
of production and product distribution. A bureaucracy was
created and co-operative officers were selected and
trained according to standard civil service criteria.
Despite the fact that the co-op officers emphasized the

fact that the co-op officers emphasized the communal character of co-operative enterprise and taught the people cute little ditties about self-help, this was really a strategy whereby the Government mobilized the population to meet its needs. As soon as the war ended and shipments from England again became reliable, co-operatives were allowed to fade away. Those who had been involved in this co-operativism from above were frustrated and/or bitter. Mr. Dowden who had risen through the civil service ranks to become Chief Cooperative Officer by the time of the proclamation of the Co-operative Republic reflected this sordid history in the following observation. "Can Consumer Co-operatives succeed in Guyana? Up to the present time, it would take only those who have a great faith in the future to say 'yes.'"[15]

Strict hierarchical command and a strong civil service orientation determined the interaction and activities of co-operative officers. Indeed, the structure concentrated power to such a degree that many regarded the Commissioner of Co-operative Development as a dictator. It was within his authority to register or cancel a co-operative, to determine lending and borrowing policy, to regulate investment of funds, to approve the membership and even to call meetings. Dowden recognized that such a tightly centralized, bureaucratic structure would suffocate any genuine mass mobilization and he recommended the amendment of the law regulating co-operative development. It is extremely noteworthy that this appeal was ignored until 1976 and furthermore, that such changes as were made kept firm control within the bureaucratic structure and the person of the Chief Co-operative Officer and Minister of Economic Development. In short, the revised legislation extended the Government's control over the co-operative movement and imposed draconian penalties for non-repayment of loans. The opposition cynically noted that any small farmer or aspiring industrialist would be scared to death to utilize any of the co-operative financial institutions.[16]   The PNC of course has a different interpretation and believes ". . . that if they waited until the people themselves decided to work through the co-operative way, it may be too late . . . The Minister under whose portfolio co-operatives fall is the policy maker. He, the Minister, discusses what plans he has in mind with his technicians, but the plan is his, and this he gives to the Chief Co-operative Officer for implementation."[17]

This failure to invigorate and transform the co-operative sector is particularly disturbing since the Government has said that the co-operative will be the foundation of socialism in Guyana. Two years after the adoption of co-operativism, the Government still had not specified which activities would be reserved for the sector; moreover, the National Co-operative Bank operated along the lines of an ordinary commercial bank.[18] Dissatisfaction with the mixed signals emanating from the Government in regard to its new development strategy grew more intense. As one local newspaper observed, "This apparent absence of detailed planning is a factor that has characterized a number of government programs, especially in the past six months or so. As a result, many of them have foundered soon after birth, and will only be remembered by the slogans they bore, and which seem to be more important than the undertaking itself."[19] Nor was the skepticism relieved by the long delayed publication of the 1972-76 Development Plan in late spring 1973. It was originally intended that this plan for tri-sectoral development would cover the 1971-80 period. In the after-glow of Republic status a rush of development planning activities commenced. Government planners were sent to elicit grassroots inputs and to explain the Government's new initiative. Unfortunately, this attempt at grassroots democracy was brought to an abrupt halt in mid-1970 when the mass mobilization proved to be "politically explosive." The chief economic planner, Wilfred David, was run out of the country.[20] Henceforth, a more modest approach and time frame were adopted. Obviously, the PNC was trying to devise a top-down model of socialist transformation with guarantees that no real power slipped into the hands of the people. The PNC elite was to maintain complete control.

The most common reaction to the Government's long awaited explanation of its new development thrust was incredulity. One reviewer after another questioned the plausibility of the plan which said it sought to make the co-operative the vanguard of the socialist revolution in Guyana while alloting it a mere 10 per cent of the projected capital expenditure.[21] Even more disturbing was the Government's seemingly deliberate intention to blur the distinction between the co-operative and private sector. In a prepared address to the Georgetown Chamber of Commerce, the Minister of Economic Development and one of the key architects of the new plan explicitly stated that the private sector included the co-operative sector. Furthermore, when the Managing Director of the Guyana National Co-operative Bank was asked to comment on this apparent change of policy, he said that he welcomed the new definition since he could not envisage the co-operative as being distinct from the private sector.

Shortly thereafter, however, the Government explained that this was not really what the Minister had meant to say.[22] Nonetheless, if Guyana still did have a third sector, no one could figure out how it would fit in. As one critic of the new plan noted it gave no indication of the types or size of co-operatives to be developed nor where the financial resources to cover the sector's projected 10% of planned expenditure would come from. Neither did it indicate where the managerial or technical expertise would be found to make the sector competitive with the public and private sectors. In a similar vein, another critic concluded that the proposed expenditure of G$75.5M was small but nonetheless represented more than the fledging co-operative sector could possibly absorb in five years. In addition, the proposed plan offered no details concerning how workers would be included in the decision-making apparatus and thus ultimately control the economy in accordance with the PNC's professed socialist objectives.[23]

The implausible and grandiose quality of the plan did not confine itself to the co-operative sector. The plan projected a total development expenditure of $1,150M over a five year period. The Government was to find and invest $650M and the private sector (including the co-operative) was to invest $500M. During the previous five year period, government and the private sector had invested a total of $527M and this covered the honeymoon period in the relationship between the Government, local and foreign capital and the U.S. Government. Washington alone had supplied $220M.[24] With Nixon in the White House and local and foreign capital nervous over the meaning of the Government's policy shifts, a development budget of such magnitude was at best a "precipitous and over-ambitious acceleration of growth."[25] The plan was filled with contradictions. While the "socialist" development plan called upon the private sector to double its previous level of investment, and relied upon increased inflows from the international capital and aid markets, it also proclaimed the goal of ending foreign dependency. The PNC Government said its policy was to utilize labor intensive technology and local resources in order to "Feed, Clothe and House the Nation" by 1976. Trade linkages were to be re-directed toward other Caribbean and Non-Aligned countries.[26] The ambiguity of the 1972-76 Co-operative Development Plan thus only served to confirm the fears of already leery capitalist investors.

On the international front, the Government was becoming an increasingly active member of the Non-Aligned Movement which was newly focused upon economic issues.

In 1970 Guyana was the only country from the Western hemisphere to attend the Non-Aligned meeting which ratified the Lusaka Declaration calling for permanent sovereignty by developing countries over their natural resources.[27] In 1972 despite the heavy expense on an already overly strained budget, Guyana volunteered to host the Meeting of the Foreign Ministers of the Non-Aligned Movement. The Government thereby took the initiative in drawing up the meeting's agenda and thus guided the deliberations which would culminate in the next year's Summit Meeting in Algiers. It was in Algiers that the Non-Aligned nations hammered out the proposal which would become the 1974 U.N. Declaration for a New International Economic Order. They demanded that patterns of trade, investment and aid established when most Third World countries were still colonized be made more equitable. The international monetary crisis, rapid inflation, global recession and, of course, OPEC success, gave a certain quality of inevitability and divine justice to their demands.

Guyana took up the leadership of the struggle to create the NIEO in the Caribbean. On April 26th 1970, Burnham disclosed that his Government would seek 51 per cent participation in major industry either directly or through co-operatives.[28] In July, he took a regional approach to the PNC's resurrected radicalism and proposed to Jamaica and Surinam the formation of a bauxite producers' association. Guyana's neighbors had not yet joined the North-South confrontation and the cartel proposal fell on deaf ears. Indeed, Caribbean governments feared that Burnham's pugnaciousness might cause foreign capital to see the region as a less desirable place to invest.[29] Nonetheless, the Government did receive encouragement in its new determination to nationalize control of the economy. It came from a most unexpected source. In October 1970 the Canadian Cabinet indicated that it was sympathetic with the goal of Caribbean governments to gain a share in the equity and control of multinational corporations operating in their territories. The decision of the Canadian Government not to intervene in any struggle between a Canadian company and a foreign government over equity and/or control was underscored when Paul Martin, a Canadian legislator, visited Burnham in November.[30] Burnham lost no time, on November 30th he announced plans to seek controlling interest in Demba, a local subsidiary of Alcan Aluminium, Limited. The Government proposed to purchase 51 per cent of Demba's assets to be valued at written down book value and paid for out of future earnings. At the same time, the Government emphasized that it was trying to do no more than advanced countries such as Australia, Canada and Japan were doing.

Alcan was working from a different set of assumptions. Despite its government's advice to cooperate with requests for participation, the company preferred the American domino theory and countered Burnham's proposal with an outrageous set of conditions. Demba would sell 51 per cent of its shares to the government provided that (1) the government agreed beforehand to the expansion of calcinated bauxite mining and raised $50M from the World Bank for this purpose; (2) that all of Demba's assets ($100M) be transferred to a new entity as debt carrying normal commercial interest rates to be paid to Alcan; and (3) that the new company be completely exempt from all forms of taxation, royalties, imposts, everything. In brief, Alcan would gain ownership of 49 per cent of a new company worth $150M without investing a dime and receive full compensation for its existing investment![31] The feeling of national outrage was intensified when it was discovered during the negotiations for participation (which lasted just one week) that Alcan had no intention to increase alumina production or to build an aluminium smelter in the country even if cheap hydro-electric power were made available. Alcan had played Guyana for the fool. There was a public outcry for vindication which echoed into normally conservative chambers. Even the pro-capitalist United Force which still opposed national-ization, was forced to agree that ". . . no Government would accept the conditions as set out by Alcan. This is an insult to the integrity of a nation and nobody can tolerate that . . . ."[32]

Alcan had miscalculated. Its hardline approach left the Government no choice but to nationalize. Up to this nasty incident, the Government had insisted that part-icipation versus complete ownership and control of the commanding heights of industry as advocated by the PPP was the answer to Guyana's economic woes. In fact, just a few weeks before, the Minister of Finance and General Secre-tary of the PNC had shouted across the legislative cham-bers at the PPP, "Nationalize? We have no need to nation-alize. We will never nationalize."[33] Now, however, the Government was forced to seek the support of its left opposition to obtain the two-thirds majority needed to change the constitutional clause guaranteeing prompt and adequate compensation in case of nationalization of property. The PPP was only too glad to support this erosion of capitalist rights. The party's enthusiasm was so great that Jagan accepted Burnham's verbal assurances in regard to the PPP's concerns for the protection of

workers and local management rather than move to amend the
bauxite nationalization bill.   On the contrary, the PPP
entreated the Government to go further and also take over
Reynolds, the American owned bauxite company, the foreign
owned sugar industry, the banks, insurance companies and
foreign trade. Jagan argued that there could be no fence
sitting on this.   Either you were against capitalism and
imperialism or for it.   The indecisive would be crushed.
Turning to Yugoslavia for advice, as Burnham had done, was
not enough.   If this was genuinely a declaration of
war against foreign exploitation then the Government
would need the support of the Soviet bloc.[34]   This of
course was rhetorical posturing.   Jagan knew that the PNC
remained wedded to the Western camp but the PPP never
foregoes an opportunity to extol the benefits of an
Eastern alliance and the Soviet socialist system.

The PPP acted on the assumption that the Government's
increasing control of the economy would heighten internal
class contradictions and external contradictions with the
imperialist system to such a degree as to topple the
regime. Their prediction was only partially correct.   On
the one hand, the takeover was finally accomplished in
such a way as to minimize antagonism within the international financial community. On the other hand, although
internal class conflict has increased so has the Government's capacity to intimidate and economically victimize
the workers.   Fear of demotion, unwanted transfer to the
interior, firing or hopes for a promotion or raise have
stifled protest and led even PPP supporters to submit to
the discipline of the ruling party. Contrary to the PPP's
doctrinaire reading of Marx and Lenin, every state takeover of industry is not a step toward socialism. Now by
its own admission, the PPP must deal with the frightful
consequences of having naively supported the mortgaging of
the nation's resources to the international banking
community:

> . . . we cannot have an accurate idea of the
> extent and longevity of our debt to these na-
> tionalized corporations as we are not cognizant
> of the terms and conditions of these national-
> ization agreements. . .Whilst we unanimously
> support nationalization that support cannot be
> given blindly for what appears to be purchase
> agreements with foreign companies of which we
> know little . . . As a result of this reformist-
> capitalist form of nationalization our economy is
> more tied to the powerful financial monopolies. .
> .than it was when tied to the MNCs . . . .[35]

The recognition came too late. When the Government needed the support of the PPP it was given uncritically for the opportunity to prove its radicalism. This is part of the reason why many Guyanese respond with contempt or amusement to the PPP's claim to the vanguard role in the country's struggle for socialism.

Of course the PNC never intended that foreign and local investors should take its fiery speeches and sabre rattling seriously. While bearing its teeth for the crowd, the Government desperately sought to reassure its traditional sources of development capital. Arthur Goldberg, legal consul to the Reynolds Company and friend of Forbes Burnham, was asked to mediate the nationalization negotiations. Guyana, Goldberg pointed out, had to think about its reputation in the international capital and aid markets.[36] The Government decided to follow Goldberg's advice. Demba would be paid more than the written down book value for its assets and the payment would be insured against the Government's general revenues instead of future earnings. As to the future, Burnham promised that nationalization would only be undertaken if all discussion aimed at reasonable participation failed. To prevent any possible misunderstanding, the Foreign Minister was sent abroad to reiterate Guyana's commitment to seeking foreign private capital. On November 31, 1972, the New York Times quoted S. Ramphal as saying, "We have got to disabuse the investors' minds about Alcan. Enlightened businessmen should be more happy than fearful at what we are doing . . . . What we want are relationships on an agreed partnership. This is what provides security for the investors."[37] The Government's reassuring gestures were not without effect. Even before the generous terms of compensation were announced on July 15, 1971, the Chase Bank agreed to loan the new state owned corporation $8 million. This was quickly followed by a $10 million loan from the U.S. dominated World Bank.[38] The U.S. did not approve the new leftist orientation of the PNC but the State Department did not make the mistake of seeing in Guyana another Cuba. On the contrary, throughout the 1970s when PNC radicalism reached its zenith, the U.S. remained convinced that the PPP, with membership in the world communist movement, represented the real threat to U.S. vital interests in the country.[39]

Nonetheless, the balance of political forces within the PNC and the nation was slipping to the advantage of the leftist forces. The Young Socialist Movement (the youth branch of the PNC) had called for the nationalization of the sugar industry, trade with Cuba and increased diplomatic ties with the Socialist world as early as 1967. ASCRIA (the cultural arm of the PNC) had pushed for the nationalization of Demba because of its apartheid system of discrimination within the company and mining community. Radical intellectuals like those who belonged to the Ratoon Group published works which insisted that the only way to develop Guyana would be to end the stranglehold of foreign capital. In this context and in the face of mounting economic crises, the PNC was moving to the left. No amount of temporizing or double-talk could reverse this or remove the fears within the naturally skittish private sector. This point was underscored by the United Force.

> You cannot change important principles of your Government overnight without exhausting every possible means of avoiding a shift in principles . . . people will turn around the next day and ask: what is the word of this Government worth, what is the word of the Prime Minister worth, what are these things I have in writing worth? Absolutely nothing![40]

Nationalization had been politically easy but this was precisely what was going to make it difficult to contain. The PPP's call for total nationalization of the economy no longer appeared so reckless in the light of the insulting response of Alcan to the PNC's reasonable approach. Indeed, as a consequence of this acrimonious experience, the PNC leadership privately decided to take complete control of existing foreign enterprises when it deemed such a move desirable and to let the 51 per cent policy apply to future foreign ventures.[41] In sum, radical alternatives had gained increased legitimacy and the private sector was less willing to risk investments.

Like Topsy, the state sector grew. It is only later events (i.e., the PNC's conversion to Marxism-Leninism) and the efforts of party ideologues to recast history which support the claim that there was a deliberate plan at work. Uncertainty, contradiction and vicissitudes were the principal architects of policy. For example, the Government nationalized Guyana Timbers Limited in March 1972 when the private owners threatened to close it down due to heavy losses. In the words of the Minister of Works and Communication, "This Government, in order to protect the workers, acquired that enterprise and has turned it into a profitable and viable undertaking."[42] Minister Hoyte used the same rationale to explain the Government's acquisition of the Guyana Graphic Company, a daily newspaper, in 1974.[43] The Government still rejected the PPP's "tek dem all" policy despite the fact within the PNC supported a more radical approach to nationalization. Furthermore, through the offices of the Guyana Development Corporation and foreign embassies, the Government still courted foreign investment. This point bears emphasis. Let us quote Minister Hoyte more extensively.

> . . . under the CARICOM arrangement, there is a common incentive applicable to all the countries that are called the more developed countries of CARICOM, Guyana, Barbados, Trinidad and Tobago and Jamaica. These incentives, sir, go into several years of income tax holidays, nine years, in fact, where those companies qualify by showing that they use local material and local labor . . . Up to the time this Government took office, five years tax holiday was offered, this Government has made it nine years.
>
> When we took office, the duty on machinery was as high in some cases as 10 per cent. Today, it is no more than 2 or 3 per cent. As a matter of fact, invariably, any business firm coming up to purchase machines for extension or new installation can, and do in fact, import its machines free of duty . . . .
>
> Let us look at other aspects. Let us look at the rate of interest which companies which want to produce are required to pay. In Guyana, the rate is one of the lowest in any part of the world. . .[44]

Firstly, these comments were made by a prominent Minister who was rapidly gaining a reputation for being one of the ideologues for the PNC's new Marxist-Leninist orientation.[45] Secondly, this recitation of a quintessential capitalist development strategy was made in October 1974, just two months prior to the PNC's Declaration of Sophia which is regarded as the party's blueprint for socialist revolution in Guyana. We will return to this glaring inconsistency briefly but first let us examine the emergence of the Government's increasing control of foreign trade.

In July 1970, the External Trade Bureau was formed to attempt to deal with the country's deteriorating balance of payments position. The ETB took control of the importation of many items (particularly those which could be obtained cheaper from socialist countries) and distributed them to retailers who were only allowed a fixed mark up. The local bourgeoisie, almost totally concentrated in the distributive sector, saw the move as a declaration of class war.[46] Bookers, the foreign owner of the largest department store and supermarket in Guyana, saw the Government's foray into import and distribution as a minor irritant but an ominous limitation of its commercial freedom of action.[47] In order to save scarce foreign exchange, the ETB prohibited or restricted the import of hundreds of items which were deemed luxuries, non-essentials or products for which local substitutes could be found. Many more items could be imported only under license. Imported consumer favorites such as raisins, sardines, onions, garlic, white potatoes and canned juices disappeared from the shelves only to be found on the exorbitant black market. The middle classes found it almost impossible to buy new cars, stereos, refrigerators or other fancy appliances.

Despite the howls of protest from all quarters, the Government's policy was not a deliberate attack aimed at anyone's class interests. According to Fred Sukdeo, a Guyanese economist, the policy reflected a series of ad hoc decisions.[48] The policy immediately suffered from the absence of sufficient and qualified staff. The goal was not to strangle the private sector. Additional restrictions on imports, foreign travel, etc., were imposed only as required by an increasing foreign exchange crisis. Likewise, the acceleration of trade with the Sino-Soviet bloc was mostly symbolic. In 1973, only 3.4% of Guyana's foreign trade was with the Socialist countries.[49] The ETB did not become self-supporting until 1972. Nevertheless, private investors responded by drawing in their purse strings.

Repeated shocks emanating from the world capitalist market drained the Guyanese economy of its limited vitality. Three years after the decision to nationalize Demba, Burnham used the same occasion, the Republic Day Celebrations, to warn the nation that "Today, the enemy is the collapse of our economy such as we are witnessing in other lands . . . ."[50] He was not exaggerating. The Government's import substitution policy was failing. Severe rains had seriously cut sugar, rice and bauxite production and thereby reduced the earnings of critically needed foreign exchange. At the end of 1973, Guyana's foreign exchange reverses reached an historic low of $39 million.[51] The oil crisis aggravated the situation by adding another $75 million to the import bill. The balance of payments deficit widened dangerously and credit was increasingly difficult to find. Loans on concessionary terms were virtually non-existent and poor countries were being forced to compete with industrialized nations which

were also fighting against oil fueled deficits for commercial loans on the Euro-dollar market.[52] For Guyana,
the situation was quickly growing out of control despite
the Government's attempt to gain better control of
the economy since 1970.

Ideological Subterfuge and Authoritarian Restructuring

Previously submerged authoritarian tendencies within
the PNC Government took control. On the economic front,
the Government had lost most of the support of its neo-
colonial allies. The former alliance with the local
bourgeoisie was completely shattered and the relationship
with international capital was seriously strained.   The
amount of direct U.S. investments was falling;[53] no new
major aid or loan packages were granted by the U.S. under
the Nixon Administration;[54] the Reynolds bauxite subsidiary was "acid-stripping" its Guyana operations in
preparation for nationalization[55] and Bookers while
continuing its expansion would only invest locally generated funds. On the political front, opposition to PNC
rule was growing. In 1974 a group of young Black and
Indian Marxist intellectuals organized the Working Peoples' Alliance. Although not at first a political party,
its potential consequences were tremendous and readily
apparent. The organization was popular among the youth
and successfully organizing in the urban areas and
bauxite communities which were formerly PNC strongholds.
This disaffected mass base had stayed clear of the PPP for
capable of mobilizing this well of Black discontent.
National elections which were scheduled for July
1973 forced the Government to deal with its deeply eroded
political position head-on. The PNC fell back on familiar
tactics and rigged the elections. This time however, the
party gave itself the two-thirds majority needed to make
sweeping constitutional and policy changes. Moreover, the
PNC declared that its overwhelming electoral victory was a
popular mandate to proceed with a socialist transformation
of society. M.F. Singh, leader of the United Force, was
outraged by the impudence of the claim.

> What utter nonsense! When did they do this?
> How did they do this: Socialism was never an
> issue at the 1973 'selections.' It was after
> the PNC gave itself the two-thirds majority in
> July 1973, that the PNC started forcing socialism
> down the throats of the Guyanese people . . . we
> all know that the PNC was the Party which in 1963

> and 1964 paraded the streets of Georgetown with the United Force protesting against communism . . . Cuba, . . . Russia, and the PPP friendship with Cuba and Russia.[56]

The points raised by Mr. Singh are important. First, Guyanese have never really registered their opinion on the choice between capitalist and socialist development. When Blacks and Indians overwhelmingly supported the united PPP in the early fifties, they were expressing a desire for independence and a better life. Once party politics bifurcated along racial lines a vote for the "communist" PPP or the "socialist" PNC was in the vast majority of cases evidence of racial loyalty. With the introduction of electoral rigging in 1968, the whole concept of representation and official responsiveness became void. In sum, we have no reliable indication of either an understanding or acceptance of either a Soviet or socialist model of development on behalf of the population. It is both tragic and ironic that the PNC claims a socialist majority in the country on the basis of its support and that of the PPP.[57]

Singh's second point in regard to coercion is also significant. Up to 1974, the PNC had used all sorts of devious, illegal means to exercise state power. After the elections of 1973 however, the party openly adopted repressive measures to maintain its rule. In November 1973 the Prime Minister declared that the PNC ". . . had become the major national insitution and that the Government was . . . merely one of its executive arms."[58] The PNC's declaration of party paramountcy was a maneuver designed to strengthen the position of an unpopular, crisis ridden regime. It was not socialism but rather the PNC which was being shoved down the throats of Guyanese. At the same time, the Government was offering the most lucrative investment incentives to date to foreign and local capitalists.[59] With party paramountcy, civil servants, including members of the police, were forced to pledge loyalty to the PNC rather than the Constitution or the President of the Republic.[60] The reaction was predictable, according to Mr. Singh,

> I have been talking to civil servants and already they are very worried. Some, of course, have jumped on the bandwagon and they are drifting along with the tide, in fact, getting very much involved. Others, however, are afraid. They have been forced or coerced into attending indoctrination courses . . . They have been asked

> to take part in political activities and to
> attend rallies and marches and so on and, out of
> fear for their future, fear for their jobs, some
> of them have in fact, just taken the easy way out
> and have done what they were asked to do.[61]

Burnham did not deny what was going on other than to note
his preference for the term "educational courses" instead
of "indoctrination." The concept of the impartiality of
the civil servant was, he pointed out, "utter nonsense . .
. no one is being asked to sell his conscience . . . this
Government has seen it fit to expose senior public servants to the rationale behind the Government's thrust and
development program, has sought to expose them to what
informs the Government, the PNC, in ideological terms."[62]

In December 1974, at a special Party Congress,
Burnham declared that the party would be radically restructured to meet the demands of Guyana's new situation.
According to Burnham, "the hanger-on, the bandwagoner, the
opportunist, the lukewarm" would have to be eliminated.
The PNC had, Burnham warned, to prepare itself for battle,
". . . the Party must be highly organized and well-trained
to mobilize the nation, give leadership to the people and
identify and rout the enemy regardless of the guise in
which he clothes himself."[63]   This, Burnham exclaimed,
was the minimum necessary to fulfill the party's new
commitment to socialist revolution.   He also announced
that it was time to re-write the Constitution so as to
reflect the new socialist policy.   Burnham's long reputed
political acumen was in evidence.   Faced with the task of
repressing its growing opposition, the PNC opportunistically adopted Lenin's prescription for correct party
organization.   The move was brilliant.   Marxism-leninism
provided a near perfect ideological subterfuge for the
authoritarian restructuring of the ruling party and the
state apparatus.   Both the PPP and WPA, the two major
opposition forces in the country, adhere to the principles
of marxism-leninism.   Thus the PNC succeeded in expanding
its coercive capability while claiming to implement the
program advocated by its opposition.   While not completely
disarmed by this maneuver, the opposition did find itself
in the awkward position of trying to explain the difference between repression and democratic centralism and
other fine points of theory.

In addition to political opposition, the PNC regime
had to deal with the increasingly restive labor movement.
The corrupted leadership of the TUM was rapidly losing
control over its membership.   The PPP retained firm
control over the sugar workers who would strike the

industry whenever their political-trade union leaders deemed it necessary. More ominous were the emerging likages between the bauxite workers once loyal to the PNC and the WPA movement. As the general economic crisis mounted, the Government could not tolerate the endemic production stoppages.

The PNC cleverly sought to exploit the radical shift in its ideological posture to extend its control over the labor movement. The year 1972 marked the beginning of the PNC's major push to restructure the TUM. In January, the renegade leadership of the Trade Union Congress (TUC) tried to convince the Government of its indispensability amidst the circulation of rumors in "high circles" that trade unions were an unnecessary hindrance to progress at Guyana's current stage of development.[64] At the TUC's 19th Annual Conference in November the delegates endorsed the emergent leftist orientation of the ruling party and passed a resolution the Government later called its "mandate" to take control of the economy. Subsequently, the Government reconsidered its veiled threat to destroy organized labor and instead spelt out the new terms of survival to the Congress' leadership.

According to the ruling party, the declaration that Guyana was now a co-operative Socialist Republic represented nothing short of a revolution.[65] Indeed, Desmond Hoyte, Minister of Works and Communication, claimed that the PNC's adoption of a socialist path of development and the Party's acceptance of the TUC's resolution calling for the nationalization of major industries represented the end of class struggle in Guyana. The working classes had won. The PNC was determined to develop the country in the interests of workers-- now the hegemonic class in society. Of course, this revolutionary development would necessitate the transformation of the consciousness of workers, particularly organized labor. The old ideology of inevitable conflict between capital and labor and the associated activities of go-slow, sabotage, strike and the like would have to be replaced, he argued, by an attitude of cooperation as Government increasingly became the owner of industry. Minister Hoyte was insistent on this point, "Unions cannot call upon the Government to organize the economy in the interest of the people and, at the same time, reserve the right to damage the economy by negative policies. The very call by the Trade Unions for Government to socialize the economy is a call for partnership."[66]

The Government's vision of this new partnership between the state and organized labor is worthy of careful attention. There were stupendous promises of increased power for the TUM. The Government proclaimed that trade unions should take responsibility for implementing a system of workers' self-management in the economic sectors where they operated. Workers, through their unions, were promised the opportunity to make management and investment decisions, set production targets, fix and enforce occupational health and safety standards and to prescribe fair remuneration for workers within formulas and guidelines established at the national level. In sum, Government argued that its ultimate goal was "the eventual transfer of economic power within the state to the workers."[67] The oppressive wage nexus was to be smashed and workers' passivity transformed into self-management. In the private sector, the content and social significance of ownership was to be changed through a legal process which would progressively eliminate exploitation. After reviewing official statements and documents pertaining to the Government's new, socialist labor policy, one must be impressed by the skill with which they articulate the principles of Humanist-Marxism associated with the Yugoslav model of workers' self-management. This is not however surprising since Burnham has long admired Tito's foreign and domestic policies. The problem in Guyana is not therefore one of theoretical impoverishment but rather the will and capacity to implement.

Workers' self-management supposedly represented only one aspect of the proposed vast extension of democratic practice. Minister Hoyte also promised that the haphazard pattern of participation in national decision-making by trade union leaders would be strengthened and institutionalized. Currently, according to Minister Hoyte, "The movement is represented on Boards, Committees and Commissions and, through the TUC has consultative and advisory status with the central government. Trade Union leaders have direct access to the Prime Minister and all Government Ministers and, from time to time, are briefed on current issues of national importance."[68] Therefore we find Minister Hoyte's conclusion that the system is "almost casual and wholly unsatisfactory" very curious. Indeed it is wise to look this gift horse in the mouth.

The role of trade union leaders as national leaders was to be institutionalized; but, in exchange, they would have to give up their old preoccupation with wage and fringe benefits increases. In a socialist Guyana, they were admonished to concentrate on securing the interests of their membership through the development of the entire country. There is nothing inherently nefarious about such a proposition. However, in the case of Guyana, it has consistently been used as an excuse for denying workers an improved standard of living while allowing politicians and bureaucrats to squander the limited national surplus on luxury consumption and inept purchases of capital goods.

According to the ruling party, the goal of national economic development cannot be accomplished through traditional trade union tactics of confrontation. Instead of assuming a reactive stance towards national policy, the TUM was urged to submit "carefully researched and documented" recommendations. Where the unions would obtain the financial and technical resources this type of policy planning would entail was not specified. Nevertheless, once the TUC was given this privilege of expanded participation in national decision-making,

> . . . it would be an impossible situation, if thereafter, individual unions reserved the right to resort to industrial action which would impede or frustrate the attainment of those objectives. The central organization of the trade union movement must be structured to enable it to speak authoritatively on behalf of the whole movement and to enter, on behalf of the movement, into binding agreements on national objectives and strategies. Once such agreement has been arrived at neither side must be free to renege.[69]

Stripped of its ideological and rhetorical flurries, the PNC policy of partnership between labor and government boils down to a hard-line attempt to control labor's defiance and eliminate production stoppages. With trade union leaders formally included in the national decision-making process, the Government claims legitimacy for its charge that any industrial action against a state-owned enterprise represents political subversion. Those guilty of mounting a "political strike" cannot, the Government warns, expect to be treated as ordinary trade union leaders or strikers. Thus people engaged in industrial action over such issues as wage increases, profit-sharing and pension funds, have been ipso facto treated as enemies of the state. With this miracle of definition accomplished, the PNC regime has resorted to mass firings, arrest and imprisonment of striking workers and their leaders, used the police and army to intimidate and brutalize strikers and recruited scab laborers by the thousands to crush "political strikes."

The Organization of Working People of Linden, a group linked to the WPA has charged that:

> The petty bourgeoisie of the PNC and at GUYBAU
> are afraid of the workers and at the same time
> contemptuous of them . . . The Chief Executive
> Officer of GUYBAU, who is part of the PNC leader-
> ship, has time and again made it known that his
> vision of workers' participation must not inter-
> fere with the pattern of authority at GUYBAU.[70]

Moreover, according to OWPL/WPA, labor-management
relations did not improve, let alone undergo a revolutionary transformation, following the nationalization of
Alcan.

> The structure of day-to-day running of GUYBAU
> made it even clearer that the essential condition
> of labor continued unchanged . . . . One of the
> most striking facts is that the allocation of
> surplus within the industry as between workers
> and management is no better and perhaps worse
> than it was previously . . . the wage/salary
> allocation made a mockery of the claim that top
> management are simply "workers" . . . the
> similarity in prevailing conditions cause workers
> to refer to the Guyanese elite in the industry as
> "the new Canadians."[71]

Thus for workers in Guyana as in most other state socialist systems, the creation of a "workers' state" has meant
more, not less, discipline imposed by capital.

World Market Impetus to PNC Radicalism

Not withstanding the ideological posturing, the sharp
radicalization of the PNC's economic development policy
was primarily caused by a dramatic reversal in the world
capitalist market. In 1974, market forces which normally
operate to the detriment of Third World nations moment-
arily gave the advantage to the underdog. The PNC Government moved quickly to exploit the temporary advantage. In
January 1974 it intervened in the marketing of sugar to
ensure that it was sold in the most lucrative markets; in
March, Guyana attended a meeting called by the Jamaican
Government to discuss the formation of a bauxite producers' cartel. In June, the PNC enacted an export tax on
the windfall profits of sugar. In August, the Government
joined the newly formed International Bauxite Association.
In September, the Government imposed a levy on bauxite
exports. The PNC was riding high. Burnham with false

modesty proclaimed, "We have been proved right.   This I observe without beating of drums and boasting."[72]   He was referring to his Government's long standing support of producer cartels.   In November, Guyana became a founder-member of the Group of Latin American and Caribbean Sugar Exporting Countries.   Like other Third World governments, the PNC regime thought that it had finally come up with the means to obtain ". . . a greater and reasonable share of the profits which flow from the exploitation of natural resources."[73]   Burnham boasted that "The receipts from sales by Guybau boosted our foreign reserves, and the enhanced price of its products produced profits for the nation's use.   Both the reserves and the profits would have gone elsewhere if Guybau were not nationally owned."[74]   For the first time the dream of independent economic development appeared within reach.   Prominent members of the PNC felt they would no longer have to go hat in hand to the developed world begging for investment funds and aid.

The Declaration of Sophia promulgated at the December 1974 Party Congress exuded the PNC's new sense of confidence and defiance.   Within the Government it is considered the PNC's socialist manifesto.   In sum the state sector was assigned the responsibility for most investment but the private sector was not eliminated.   According to the Declaration, trisectoral development was still the backbone of the PNC's approach.   As in the past, details were upstaged by grandiose and vague promises.   Burnham said that the Government would proceed with its policy of local ownership and control of natural resources.   The activities of foreign banks would be curtailed.   In regard to both no time frame was provided. A fuller statement on the proposed role of the private sector was forthcoming, Burnham promised, and it would provide a guarantee against confiscation.   For the time being private investment was still welcome so long as the state or co-operative sector held majority equity.

The year 1975 represents a watershed in PNC radicalism.   On the one hand, the U.S. had cut off all aid and loans to underscore its displeasure with the PNC's increasingly progressive foreign policy--particularly its support of Cuba and the marxist-leninist liberation movements in Southern Africa.[75]   On the other hand, the Government was fed up with the niggardly investments of the private sector even after a most solicitous courtship.[76]   And, with foreign exchange reserves comfortably above $200 million, it was prepared to undertake the necessary investments itself.   The Minister of Finance announced that new lending guidelines would favor the public and co-operative sectors and that commercial bank credit would be severely restricted for private investors.[77]   In May, the Parliament enacted legislation to facilitate the acquisition of privately held property.[78] In December, the PNC used its fraudulent two-thirds majority to overide the strenuous opposition of the United Force to a Constitutional amendment which virtually removed all safeguards against expropriation.[79]   It was clear that more nationalizations were anticipated.

Nevertheless, the PNC did not allow its new radicalism to stand in the way of expediency.   Pragmatism and moderation remained the cornerstones of PNC strategy.   The U.S. bauxite subsidiary was nationalized in January 1975 only after Reynolds refused to pay a production levy similar to one the company had agreed to pay the Jamaican Government.   How could the Government, the PNC asked, tolerate this affront when underdeveloped economies like Guyana's were being destroyed by inflation and Reynolds was reporting record profits?[80]   Even the conservative U.F. felt that the Government had been "most reasonable" in its handling of the situation.   According to a local newspaper, the Government displayed ". . . monumental patience and moderation in dealing with a corporation whose ideas of mining development revert to the last decades of the 19th century when Cecil Rhodes and his group were exploiting the Kimberly diamond and gold mines of South Africa."[81]   As in the Alcan case, the Government was forced to nationalize the expatriate enterprise in defense of sovereign right and national dignity.

Such negative experiences with foreign capitalists strengthened the position of the radicals within the PNC. Negotiations for possible joint ventures between the Guyanaese and Cuban Governments were undertaken.   Burnham made an official visit to Communist China.   Finally, by the time of the PNC Party Congress in August 1975, it appeared as if the ruling party was ready to renounce moderation and its historic alliance with the West.   The PNC formally adopted marxism-leninism as its official ideology.

NOTES

1. Quoted in Cheddi Jagan, The West on Trial: The Fight
    for Guyana's Freedom, rev. 2d ed. (Berlin, German
    Democratic Republic: Seven Seas Books, 1975), p.
    368.
2. Parliament House, "Debates," Georgetown, Guyana, 21
    June 1966. (Typewritten.)
3. W. Arthur Lewis, "The Industrialization of the
    British West Indies," reprint from Caribbean Economic
    Review 1 (December 1949) and 2 (May 1950): p. 55.
4. Kempe Hope, Development Policy in Guyana: Planning,
    Finance, and Administration (Boulder, Colorado: West-
    view Press, 1979), p. 114.
5. Charles Jacob, Jr., "Guyana: Victim of Electoral
    Fraud," Political Affairs 48 (May 1968): 28.
6. Rickey Singh, "909 Strikes, $7 Million Lost in Wages
    in 6 Years: Which Way Towards INdustrial Peace," The
    Sunday Graphic, 11 July 1971.
7. As of December 1980 the PNC Government still had not
    made these minimal changes in the country's indus-
    trial relations framework. Interview with Norman
    Semple, Chief Labor Officer, Ministry of Labor; also
    see Harold Lutchman, "Trade Unions and Human Rights,"
    n.p., speech delivered on August 16, 1980.
8. Charles Jacob, Jr., "Guyana: Victim of Electoral
    Fraud," Political Affairs 48 (May 1969): 23-28.
9. Juan Linz, "Authoritarian Regime: Spain," in Reader
    in Political Sociology, ed. Frank Lindenfeld (New
    York: Funk and Wagnalls, 1968), pp. 129-148.
10. Forbes Burnham, "Declaration of Sophia," Address by
    the Leader of the PNC at a Special Congress, Pln.
    Sophia, Georgetown, December 1974, pp. 7-10.
11. Hugh W. L. Payne, "Historical Background to Co-
    Operative Socialism in Guyana," n.p., 19 January
    1975, National Archives, Georgetown, Guyana.
12. Quoted in Payne, ibid.
13. Harold Lutchman, "The Co-Operative Republic of Guy-
    ana," Caribbean Studies 10 (October 1970): 114-115.
14. Ibid., p. 109.
15. Rubert A. Dowden, "The Co-Operative Movement in
    Guyana," n.d., n.p., National Library of Guyana,
    Georgetown, Guyana, p. 33.

16. Parliament House, "Debates," Georgetown, Guyana, 23 June 1976; The Chronicle (Georgetown), 13 June 1976.
17. Dowden, pp. 52 and 56.
18. The Guyana Graphic, 19 March 1972.
19. Ibid.
20. Wildred David, "Comprehensive Planning for National Economic Development: The Guyana Experience," in Studies in Postcolonial Society, ed. Aubrey Armstrong (Yaounde, Cameroon: African World Press, 1975) p. 289.
21. J. E. Greene, "The Politics of Economic Planning in Guyana," Social and Economic Studies 23 (June 1974): 186-203; Guy Standing and Richard Szal, Poverty and Basic Needs, Evidence From Guyana and the Philippines (Geneva: International Labor Office, 1979), pp. 78-79.
22. The Guyana Graphic, ibid.
23. Hope, pp. 125-133.
24. The Guyana Graphic, ibid.
25. David, p. 284.
26. Hope, p. 125.
27. Isiah A. Litvak and Cristopher J. Maule, "Nationalization in the Caribbean Bauxite Industry," International Affairs 51 (January 1975): 43.
28. The Weekend Post and Sunday Argosy, (Georgetown, Guyana) 26 April 1970.
29. The Chronicle, (Georgetown, Guyana) 23 February 1974.
30. Litvak and Maule, p. 52.
31. "Debates," 27 February 1971.
32. Ibid.
33. Ibid.
34. Ibid.
35. Peoples' Progressive Party, "Nationalization and the New Investment Strategy," n.d., n.p., available at Freedom House, Georgetown, Guyana.
36. Litvak and Maule, p. 54.
37. Quoted in Cheddi Jagan, "Guyana: A Reply to Critics," Monthly Review 29 (September 1977): 41.
38. Jagan, The West on Trial, p. 407.
39. John Bartlow Martin, U.S. Policy in the Caribbean (New York: Westview Press, 1978), p. 138.
40. "Debates," ibid.
41. Interview with Harold Davis, Guyana Sugar Corporation, Georgetown, Guyana, 24 November 1980.

42. "Debates," 23 October 1974 and 26 February 1976.
43. "Debates," 1 November 1974.
44. "Debates," 23 October 1974.
45. Given the sudden nature of the Marxist-Leninist departure, many within the Government and party were caught unprepared to speak the language of revolution let alone expound upon its deeper philosophical and historical roots. At the Special Party Congress held in December 1974, the Prime Minister and Leader of the Party announced that classes and literature would soon be made available to explain the PNC's new socialist position.
46. "Debates," 23 October 1974.
47. Booker McConnell Limited, Annual Report and Accounts 1970.
48. Fred Sukdeo, "Perspectives of Cooperative Socialism in the Sugar Industry," in Armstrong, p. 23.
49. Ministry of Information and Culture, "Guyana, A Decade of Progress," December 1974, pamphlet, pp. 50-51.
50. The Chronicle (Georgetown) 23 November 1974.
51. The Chronicle, 30 March 1975.
52. Budget Speech 1975.
53. Hilbourne A. Watson, "The Political Economy of U.S.- Caribbean Relations," The Black Scholar 11 (January/February 1980): 36.
54. Martin, p. 138.
55. "Debates," 25 September and 21 May 1975.
56. "Debates," 17 November 1977.
57. Ibid.
58. Burnham, "Declaration of Sophia," p. 11.
59. Budget Speech 1974.
60. "Debates," 25 September and 21 May 1975.
61. "Debates," 19 December 1974.
62. Ibid.
63. Burnham, "Declaration of Sophia," p. 31.
64. George DePena, "As It Is," n.d., TUC pamphlet, articles first written for broadcast on Radio Demarara, p. 31.
65. In the following discussion we use the PNC's policy pronouncements as the basis for evaluating the regime's innovation in the area of labor relations. Thus the extent of the Government's success or failure is in direct relation to the breath of its stated objectives.
66. Desmond Hoyte, "Trade Unionism and the State in Partnership or in Conflict," Critchlow Lectures Series, Publication No. 3 (August, 1973), p. 62.

67. Untitled document in the file of the Labor Code Commission, Georgetown, Guyana.
68. Hoyte, p. 57.
69. Ibid., p. 60.
70. Organization of Working People of Linden and the Working Peoples' Alliance, "The Peoples' National Congress Versus the Bauxite Workers," n.p. (January, 1977), p. 13.
71. Ibid., pp. 10-11.
72. Burnham, ibid., p 24.
73. Ibid.
74. Ibid., p. 37.
75. The Trinidad Guardian, 13 March 1977.
76. "Debates," 23 October 1974 and 26 February 1976.
77. Budget Speech 1976.
78. "Debates," 23 May 1975.
79. "Debates," 18 December 1975.
80. "Debates," 25 September 1974.
81. The Sunday Graphic, (Georgetown), 29 September 1974.

CHAPTER 5

From Neocolonialism to State Capitalism
in the Sugar Industry

The sharp radicalization of the Government's development strategy and ideology was greeted with enthusiasm by many at home and abroad. In particular, the PNC's conversion to marxism-leninism and its increasing ties to the Soviet Socialist bloc delighted orthodox marxists who believed that the quasi-socialist party had finally seen the light. Moreover, the adoption of marxism-leninism by the ruling party suggested that the ideological rift between the PNC and the PPP which contributed to the racial bifurcation of the nationalist movement in the fifties had at long last been bridged. On the surface, everything seemed to be pointing towards the reunification of the working class under the leadership of a communist vanguard drawn from the PNC and the PPP.

Jay Mandle's analysis of the nationalization of the Guyanese sugar industry strongly reflects this mood of revolutionary optimism. For Mandle the Government's takeover of the sugar industry was the logical and inevitable expression of the PNC's new radicalism. Unfortunately, his impatience to see socialism triumph in the Third World has blinded him and many other leftists to the existence of some unpleasant yet undeniable facts. First, events within the world capitalist market (obviously beyond the control of national leaders) continue to dictate the course of political and economic development in the Third World even after a country adopts a noncapitalist path of development. Secondly, socialist declarations are not necessarily related to a policy of revolutionary transformation. Very often the fiery rhetoric is simply a convenient and popular cover for the extension of state power in the economy and society.

Jay Mandle's analysis of the nationalization of the Bookers' property in Guyana misses these crucial points.[1] He completely overlooks the abrupt price explosion in the world sugar market and therefore does not see any relationship between this event and the sudden disruption of the Government's policy in the sugar industry. Instead, he naively accepts the radicalization of the PNC's ideology as the cause of its policy reversal. This revolutionary awakening is, in Mandle's opinion, responsible for the rapproachment with the country's established marxist-leninist party and the decision to nationalize Bookers' holdings. The recogniztion of GAWU is then a "quid pro quo" for the "critical support" the PPP gave to the PNC in August 1975. More importantly, according to Mandle, the Government had to have control over Bookers'

vast agricultural holding if it was to be able to "rationalize" Guyana's agricultural policy and to "integrate" the
national economy.

A close examination of the events preceding and sur-
rounding the PNC's nationalization of the sugar industry
does not bear out Mandle's analysis. Our reading of these
events indicates that the PNC used socialist rhetoric and
the banner of national interest to obfuscate its drive
to establish a model of state capitalist control in the
industry. The ideological convergence of the PNC and PPP
played only a minor part in the Government's decision to
take sugar. Far more significant was (1) the PNC's
greatly increased self-confidence in part brought about by
the commodity price boom, (2) its successful takeover of
the bauxite industry and (3) Burnham's wager that he could
out-maneuver Jagan even on his own turf. In the pages
which follow, we will trace the development of the PNC's
policy in the sugar industry. We start with the Government's neo-colonial policy in the wake of independence and
trace the steady deteriorization of the partnership with
Bookers which ended with nationalization.

Burnham's "De-Colonialization" of the Sugar Industry

By virtue of Bookers' positive image within the
Black community, the privileged status of several PNC
leaders in the company's hierarchy, the party's opposition
to GAWU, and the economy's heavy dependence on sugar,
Burnham was ready to cooperate with Bookers. Several
years later, Bookers' Chairman proudly described the
alliance between the company and the Government in glowing
terms to the New York Sugar Club:

> Guyana alone of the countries in the West Indies
> is continuing to expand its sugar production. It
> shows what can be done when private enterprise
> and government work together for the development
> of a country's natural resources. No industry
> could have greater understanding and support than
> we have had from the government under Prime
> Minister Burnham.[2]

Shortly after assuming office, Burnham met with the
Chairman of Bookers and the Demerara Company to discuss
his "de-colonization" policy for the sugar industry. The
most remarkable feature of the proposal was the extent to
which it resembled Bookers' on-going Guyanization program.
Campbell explained the objectives of "de-colonization" in

practically the same terms he had previously used to describe Guyanization. "By this concept we all mean the fuller integration of the industry with the community, to avoid its standing out—in light of its past history—as an alien organization superimposed on British Guiana."[3] The   Governments' program was basically four pronged. It sought (1) to expedite the replacement of expatriate staff by Guyanese; (2) to expand the cultivation of cane by peasant farmers; (3) to develop a proper relationship between the sugar estates and a proposed system of local government and; (4) to acquire a minority shareholding for Government in Bookers' companies in Guyana.

The first two proposals presented no problems. Bookers was, by now, thoroughly convinced of the advantages of a total Guyanization of its staff and expansion of cane farming despite the latter's reduction of company profit.[4] Furthermore, Burnham's assurances that the company could continue its mechanization program meant that much of the loss due to cane farming could be made up elsewhere. Thus "Bookers struggled to meet the government's social goals by raising the proportion of cane bought from farmers to 10% of the total output."[5] The Government accepted Bookers' argument that peasant cane could not exceed this proportion of production without jeopardizing the international competitiveness of Guyana's sugar industry.

There is also evidence that the PNC was not enthused with the idea of cane farming in the first place. The PNC's unemployed supporters were not going to leave the cities even if attractive terms were offered to prospective cane farmers. Therefore, when the National Cane Farming Committee was established in 1965, it was allocated a mere $4,182. In 1969, the sum was increased to $5,000 and in 1970, the year Guyana became a Co-operative Socialist Republic, the Committee budget rose to $25,000.[6] Cane farmers, like sugar workers, are PPP supporters; thus, the Government is not eager to increase their number or organizational strength. In 1970, the Guyana Cane Farmers' Association, which represents the majority of cane farmers and is affiliated with the PPP, was excluded from the membership of the National Cane Farmers Committee. Moreover, members of that Committee were "hand picked" by the Government.[7] Brian Scott concluded that "Bookers' interest in cane farming was social and political rather than economic."[8] It also appears that the PNC's motivation was basically public relations.

The next item on the Government's agenda, the development of local government was more ticklish. Historically, sugar planters had always opposed the development of a

system of local government. They objected to the payment
of local taxes and feared the loss of the complete autonomy they enjoyed in the countryside. Sugar was the King
and BSE preferred to foot the bill of providing community
services rather than surrender its power.[9] Thus when
the PNC   Government met with resistance it proceeded
cautiously. The dual process by which Bookers was relieved of its responsibility and authority in the sugar
workers' communities took roughly a decade to complete.
Although Burnham made the initial proposal in 1965,
Bookers did not submit to local taxation until 1970. And
it was 1972 before Bookers began to transfer legal title
to the lands in new housing areas to the government and to
make payments to the emerging local authorities from the
Sugar Industry Labor Welfare Fund (SILWF) for the provision of community services. In 1973, the local authorities received more than $73,000 from this source.[10] It
was not until 1974 as the PNC's radicalism approached its
high point, that the local authorities assumed complete
control over the lands and all civic works in communities
previously controlled by Bookers.[11] Furthermore, in
the opinion of B. Scott and the PPP, this aggressive
policy of the PNC in regard to the industry's control of
land and the special funds was not so much directed aginst
Bookers as against unorganized squatters who were seizing
the company's land.[12]

The Government's final recommendation—that it become
a minority shareholder in Bookers' Guyanese businesses—
was quite acceptable to Booker. Understanding that
Government participation would make Bookers less vulner-
able to political attack, Campbell had first proposed to
Jagan and later to Burnham that the Government should
become a junior partner in the company.[13] Furthermore,
Campbell could assure Bookers' expatriate owners that "The
Prime Minister has made it publicly plain that there is no
question of interference with the operational and technical management of the industry."[14] The London Office
has no record of why these negotiations failed, but Edgar
Readwin, then Chairman of BSE, claims they broke down
because Burnham insisted that the shares be given to the
Government. Be that as it may, the PNC Government did not
again discuss the acquisition of Bookers' shares until the
time of nationalization in 1976.

When in 1970 Burnham announced the Government's decision to seek 51% of foreign businesses, the Sugar Producers Association (SPA), clearly dominated by Bookers,
publicly responded with understanding and confidence. The
Chairman of Demerara Company cautioned against the hysteria which was sweeping the private sector.

> As we understand it, Government intends to have meaningful participation in the development of its natural resources whatever these may be and meaningful participation can take several forms to suit a variety of conditions.
>
> What I am concerned about is the way in which Government's intentions have been misunderstood and I think this has led to a widening gulf in 1970 between the private sector and Government. This is a sad state of affairs and no good to anyone. It is particularly unfortunate because it seems to me almost every private trader seems to be of the belief that Government wants 51% of his business and I know this is simply not so.
>
> On the other hand, I think some of the business community have got to wake up and realize we are operating in 1970 and not 1960.[15]

Nor did SPA panic when the Government nationalized the Canadian Bauxite Company. A confidential memo, written by a Guyanese staff member responsible for interpreting local political developments to the Bookers head office in London, coolly projected that "the sugar industry as a private enterprise is unlikely to last longer than three to five years." This prediction was made in June, 1971; Bookers was nationalized in May, 1976, exactly five years later. According to this intelligence source, Alcan had been taken first because of its repugnant colonial practices and the concentration of PNC supporters in the mining communities. Bookers had survived this long because it had shaken its historic image as the oppressive King Sugar (for the Black population at least) and could hang-on even longer by continuing to pursue its multifaceted Guyanization programs. The Government was gaining both financially and politically by leaving Bookers in control of Guyana's sugar industry. As we will shortly see, Bookers' intelligence was correct. The PNC Government did not disrupt its mutually beneficial arrangement with Bookers until external market forces significantly altered the costs-benefit ratio.

According to Bookers current Chief Executive Officer, Burnham knew "he had an efficient milk cow which he left alone."[16] Notwithstanding his increasing attraction to radical solutions to Guyana's economic problems, Michael Caine points out that Burnham remained very clear on the fact that Bookers was running the most efficient and profitable sugar industry in the Caribbean.

While Jamaica's sugar industry, dependent upon peasant cultivation, had steadily declined, Guyana's industry had flourished in the hands of expatriates. Sugar and its by-products represented roughly a third of the country's exports and contributed more than 25% to Guyana's foreign exchange earnings. In Guyana, sugar remained the "sheet anchor" of the economy.

Bookers and the Government had worked well together to expand the industry. In 1965, Bookers had 75,000 acres under cane. By 1972, 106,000 acres--all of the arable land owned by Bookers--was being cultivated. Bookers turned to the Government for assistance. Burnham did not disappoint them. Bookers' Chairman brought good news to the 1972 shareholders meeting:

> It is encouraging to be able to report that earlier this year the Prime Minister of Guyana referred to Bookers' role in Guyana as "the expert growing of sugar cane, the technologically up-to-date manufacturing of sugar, and the efficient marketing of this commodity in markets of the world in keeping with Guyana's needs and commitments." Mr. Burnham added that this Government would "insure that sugar is allowed, encouraged--indeed spurred--to expand its production and make a definitive contribution to the employment and feeding of our people and to the growth of our economy. We are prepared to give leases to Bookers for expanded cane culti- vation."[17]

Even as late as September 1975, Bookers had plans to further expand its acreage and to open a new sugar factory at Skeldon. This, in the opinion of the Minister of Economic Development, Desmond Hoyte, demonstrated Bookers' continued confidence in the economic future of Guyana and in the policies being pursued by the PNC and the Govern- ment.[18] Hoyte overstated the case; nonetheless, despite the 1971 prediction of impending nationalization and the PNC's marxist-leninist departure, Bookers did feel that it had at least another three years of life in Guyana.

Bookers and the Government verus the Sugar Workers

Clearly, Bookers and the PNC Government were agreed that output should be increased. However, greater acreage and technical efficiency are alone not enough. A coopera- tive labor force--albeit of reduced size--is still needed

for production. Bookers' post-1953 trade union policy was
the least successful aspect of its Guyanization program.
The sugar estates remained fertile fields for GAWU
and PPP activists. Bookers efforts in the sugar industry
to channel the wave of the trade union activity into a
nonpolitical direction was a total failure.[19] Thus,
despite the reforms of the fifties and sixties, Bookers
still could not control GAWU's power. More so than ever,
Bookers needed the cooperation of the Government.

Burnham's Government was happy to oblige. In the
words of a senior officer of the Ministry of Labor, "The
employers, finding themselves inadequate to deal effectively with a politically based union, continued to recognize the MPCA with the sympathetic support of the government who clearly recognize this danger of allowing this
vital industry to be controlled by its opposition."[20]
Publicly, the PNC Government played the role of neutral
conciliator. According to the Deputy Prime Minister, the
Cabinet's policy was to keep the scales balanced between
the MPCA and the GAWU.[21] Consequently, although the
Minister of Labor agreed that workers had the right to a
union of their choice, he also held that it was not the
responsibility of Government to grant such recognition.
That was a matter for the sugar employers.[22] In private, however, Burnham was quite straight-forward about
his desire to keep Bookers as a buffer between his Government and the opposition and to keep GAWU unrecognized.
"Why should he tarnish himself by directly fighting
Jagan?" observed Bookers' Chief Executive Officer. "Year
after year we faced the workers for him."[23] Indeed the
Government opposed the GAWU at one point when the SPA had
decided it would be easier and less costly to recognize
the union. This temporary rift between the SPA and the
Government developed in 1970 when the Government still
placed its political goals in the industry ahead of
economic considerations.[24] According to Norman Semple,
Guyana's Chief Labor Officer, "The MPCA had a sweet thing
going." Workers were forced to join the union and pay
dues. Those who refused were denied work on the estates.
Furthermore, if they came to the Ministry of Labor to
complain, they received only a "deaf ear." They would be
forced to go back and join the MPCA.[25] No wonder the
MPCA always supported a membership survey and offered its
membership books to the Ministry of Labor when challenged
by the GAWU. Nonetheless, the Government found it increasingly difficult to deny GAWU's demands for recognition. Despite an officially large memership and a
fat treasury, the MPCA had absolutely no control over the
labor force. Communications between the MPCA and the

workers had totally broken down and union officials could
only visit the estates with Government escorts. By 1969
it was already clear that a new policy for dealing with
strikes was needed. Burnham was ready to listen to a
controversial argument put forward by his new Minister of
Labor, Winslow Carrington. Minister Carrington, a grad-
uate of the American Institute for Free Labor Development, argued that the disruptions which were crippling
production in the sugar industry could be eliminated or at
least significantly reduced, if the GAWU was granted recognition.[26] This was a highly unusual argument within
PNC circles, yet it had obvious appeal. Hopefully, a
collective bargaining agreement would institutionalize the
conflict and compel the GAWU to observe established
procedures for handling workers' grievances. Not surprisingly, the Cabinet was not totally convinced. In
place of official bargaining status, the GAWU was granted
de facto recognition. With this stratagem, the PNC
Government sought to appease the GAWU and to preserve the
MPCA, its long time ally, as the sole official bargaining
agent for the sugar workers. In actual practice, the
Ministry of Labor became the bargaining agent for the
GAWU. GAWU would call a strike and the Ministry would
immediately intercede and demand a meeting with management
to discuss the union's grievances.[27] The Minister of
Labor and his staff would regularly "tramp through the
slushy dams and muddy swamps" to negotiate directly with
sugar workers in efforts to settle their grievances.
Harry Lall, President of the GAWU, could reach Minister
Carrington at any time.[28] For a brief period it appeared as though the new policy of working with the GAWU
had reduced the level of strike activity. However, by
1970 it was clear that the PPP was not satisfied with mere
de facto recognition.

As the dictatorial tendencies of the PNC's rule
became more pronounced, the PPP was forced to increase its
reliance upon the GAWU. In addition, the spiraling
economic crisis of the early 1970s augmented the GAWU's
economic power. Firstly, the projections for sugar prices
were very good. Secondly, the Government was in desperate
need of foreign exchange to meet its inflation driven
import bill. It appeared that "the only way that the
balance of payments can be brought into line is by an all
out effort to produce as much sugar as possible..."[29]
Thus, as the PNC began its radicalization of the country's
development strategy, the PPP prepared the sugar workers
for a "titantic battle." Jagan told workers that "they
must recognize the need to adopt new tactics and prepare
themselves to strike again and again . . . ."[30] At the

same time, a PPP spokesman   justified the party's
leadership of strike activity to the country on theoretical grounds. The party was, he contended, following the
teachings of Lenin who had taught that all class struggles
are in essence political struggles. Moreover, since
politics is the concentrated expression of economic
contradictions, trade union struggles were the first step
toward political consciousness. In sum, the PPP was quite
explicit about its intention to wage its political
attack against the PNC through the trade union struggle
within the sugar industry.[31]

It was apparent that although the Government was
prepared to accept the GAWU, it was not yet ready to deal
directly with Jagan and the PPP in the industry. Many in
Burnham's Government still agreed with Richard Ishmael,
President of MPCA. Ishmael accused the new Minister of
Labor of undermining his union and warned the Government
that his "ill-considered approach" would result in economic ruin and further political instability. "If he [the
Minister of Labor] does not know that the PPP has laid a
trap by which they hope to control the sugar industry and
bring about economic strangulation in the country, then he
is well advised to know the facts as they are."[32]   In
short, Ishmael was warning the Government that if it
followed Carrington's advice to betray its long-term
alliance with the MPCA it would open Pandora's box in the
industry. Notwithstanding, this obvious political threat,
the Government had few other options. Consequently,
the Government's efforts to curtail the production stoppages without recognizing the GAWU became more desperate
and brought it into increasing conflict not only with the
GAWU and the MPCA but the TUC and the SPA as well.

For instance, as part of its effort to maintain
its role as neutral conciliator, the Government had relied
heavily upon advisory committees and/or commissions of
inquiry.   However, the stratagem proved inadequate to
handle the escalating problems. From 1965 on, every wage
and bonus negotiation broke down necessitating the ap-
pointment of an independent body to investigate and make
recommendations.[33] The crippling recognition strike of
1970 thoroughly exposed the folly of strategy.   In an
editorial entitled "Teams and Tomfoolery" a correspondent
cynically mused that,

> Indeed the appointment of a new committee before
> the older ones have completed their tasks,
> necessarily predicates the failure of the older
> ones to do their jobs. By the same token, I may
> expect that this latest committee, by inference,

> forefated to abortion, has its successor ready on
> the assembly line . . . I am wedded without
> expectation of divorce to the belief that legis-
> lation is the only answer to the plight in which
> sugar stands.[34]

Lucian was referring to the controversial Zaman
Ali Advisory Committee appointed by Burnham during August
1970. The Low-A-Chee Arbitration Tribunal which the MPCA
and SPA had agreed to, had yet to make its recommendation.
Meanwhile, GAWU threatened to paralyze the industry until
such time as recognition was granted. Sugar workers
petitioned the Ministry of Labor to conduct a poll in the
industry. (Not surprisingly, the TUC, led by Richard
Ishmael, vehemently protested the proposed poll as a
violation of established practice and opposed the Trades
Dispute Bill as an unmitigated attack upon the entire TUM
in   Guyana.) The SPA supported the Government's labor
bill which prescribed compulsory arbitration when the
Government deemed that national interests were at stake,
but the company's said it would pay the large wage increase recommended by the Low-A-Chee Tribunal only if the
Government first agreed to increase the price of sugar
sold locally. The Government's back was against the wall.
If Carrington was wrong and they recognized the GAWU they
could lose all control over the industry. Moreover, if
the Government decided to enact its anti-strike bill, it
would forfeit the loyal support of the TUC led by Ishmael.
For the moment, the easiest move was against the SPA.
They would have to pay the wage increase without any
compensation in the form of an increased price for locally
sold sugar.

Growing Strains in the Bookers-PNC Alliance

The respite that the Government earned was limited.
The MPCA accepted the Low-A-Chee Award, but the GAWU
rejected the payments as too little and again called
the workers out on strike. The 1970 strike had already
cost the industry $12 million dollars and public pressure
was mounting to resolve the inter-union dispute once-and-
for-all.[35] The Government's position was becoming increasingly untenable. It was revealed that some members
of the supposedly independent Zaman Ali Committee were
actually PNC activists. The back room politicking which
ensued was intense. The Ali Report was submitted to the
Government on December 12, 1970. A controversy over its
contents brewed for six months. Zaman Ali, the principal

author of the report charged that Minister Carrington had
deliberately distorted the conclusions of his Committee,
"in some cases, the topic dealt with was given a completely different meaning altogether."[36] In March 1971,
he refused to present the report to the President,
saying that it suffered from "significant omissions and
changes." Nonetheless, he publicly affirmed that he was
still "an ardent PNC supporter."[37]

The most shocking discovery was the disclosure that
members of the Zaman Ali Committee were behind the feverish activities to organize a new trade union in the sugar
industry affiliated with the PNC.[38] Indeed, although
Zaman Ali originally denied any involvement in the affair,
he later emerged as the President of the Union of Agricultural and Allied Workers (UAAW). He, like the leaders of
the PPP and GAWU, also claimed that a union must be
aligned to a political party.[39] Clearly the Government
was talking out of both sides of its mouth. As it
denounced the politicalization of trade union struggles by
PPP, it dispatched PNC activists into the sugar industry
to pursue the same tactics. The Government's allies in
the industry, MPCA, the TUC and the SPA, were shocked and
outraged. Ishmael, speaking for both the MPCA and the TUC
angrily charged that "The MPCA knows that PNC activists
are trying to gain a foothold in the industry . . . If the
new union succeeds, it will be the first step towards
splitting the entire trade union movement along racial and
political lines. It could easily signify the death of
the trade union movement as an effective institution.
This is also the TUC's main concern."[40] Ishmael's
warning was, needless to say, 15 years too late. And of
course, he had played a major part in the imperialist
strategy which emasculated the trade union movement of the
fifties and institutionalized its racial and political
divisions.

The SPA also regretted the PNC's attempt to infil-
trate the industry and predicted that the already burgeon-
ing problem of labor disputes would be "aggravated more
and more by PNC-PPP conflicts in the sugar belt, mani-
fested through 'trade unions' struggles."[41] The PNC's
unexpected move led the SPA to abandon positions the
Association had once vigorously defended. To begin with,
the SPA recommended the amendment of the provision within
the labor law, which allowed a small group of agitators (a
minimum of seven) to organize a rival union and thereby
disrupt industrial relations in an entire industry.
Furthermore, the SPA suggested that it was ready to
seriously consider granting the GAWU recognition. The
rebuff to the UAAW and the PNC was obvious.

The alliance which had united the PNC and Booker was eroding. The Government's primary concern in the sugar industry was to neutralize its political opposition and it was willing to sacrifice tonnage to achieve this goal. On the other hand, Bookers' foremost concern was meeting production targets. If SPA did not meet its 1971 goal of 390,000 tons, it would not only fail to take advantage of good free market prices, it would actually suffer a loss.[42]

The Zaman Ali Report was released on June 20, 1971 and the reasons for the delay and its connection with the newly created UAAW were immediately evident. The Report made explicit that which the PNC's union organizing efforts had only suggested. The Government wanted firm control over the industry and it was willing to undermine the MPCA and Bookers to achieve it. First, the Report recommended a three year ban on the GAWU and the MPCA. During the interim, it recommended that workers be exposed to an intensive trade union educational program. Their grievances would be handled by an elected Workers' Council with district Labor Officers acting as ex-officio members. At the end of the three year ban, it recommended that the Government set up machinery to determine the workers' choice of a representational agent. Secondly, the Report proposed the creation of a permanent Sugar Board, with stationary authority to supervise conditions in the sugar industry, and prosecute breaches of its regulations and when possible, make on-the-spot arbitration decisions. Both union and management would have the right to bring complaints before the Board, but, its decisions would be legally binding on all parties. The Sugar Board would only be answerable to the Cabinet or Minister of Labor. Thirdly, the Report recommended that all estate cultivation be turned over to cooperative production by small cane farmers.[43] The Committee also encouraged the Government to press ahead in its efforts to enact the Trades Dispute Bill.

The Zaman Ali Report was a deadly challenge to the essential rights and functions of trade unions in the sugar industry and Bookers' treasured corporate freedom of action. In the extreme, both would be eliminated--unions by a Government controlled Workers' Council and Trade Dispute Bill, and private ownership and management by a Government controlled cooperative scheme. At minimum, the prerogatives of both would be circumscribed by the far-reaching powers of a permanent Sugar Board. The wave of opposition which followed in the wake of the Report's release, convinced the Government to postpone any action on its recommendations. Thus the conflicts in the sugar industry during 1970 and 1971 ended in a stalemate. Bookers remained in control of production and marketing and the MPCA retained its legal status as the sole bargaining agent for sugar workers; but, both were more dependent than ever upon a Government caught in the cross-currents of an international economic crisis, declining domestic legitimacy and the increasing radicalization of the ruling party's ideology.

Adding to the tensions, the GAWU adopted a new and successful strategy in 1972. In the past, the union had been able to paralyze production by calling industry-wide strikes; however, Bookers and the Government could always hold out longer than the poverty-striken sugar workers. Now, instead of industry-wide strikes, the GAWU kept production in disarray by closing down one estate after another for one week at a time. Now, union activists could concentrate their efforts on one estate and the financial burden of strike action could be rotated among the workers. "The technique was quite effective . . . The MPCA and the employers found it impossible to stem this tide or to mollify its effects, and the country's economy suffered a telling blow . . . ."[44] . From the Government's perspective, the cost of maintaining the status quo was mounting.

Furthermore, the growing influence of the radical wing within the ruling party made a rapprochment with the Marxist-Leninist PPP seem increasingly practical. The PPP would, of course, demand recognition for the GAWU and "at best, the GAWU can press for the nationalization of the sugar industry and force the Government to take a position on this important issue. But to judge from current trends, this does not seem to be a development so removed from the thinking of the PNC..."[45] Ricky Singh made this observation in the beginning of 1971. The appeal of the argument steadily grew as the Government successfully nationalized the Canadian Bauxite Co., strengthened its control over the TUC through its new "partnership" with labor and signed trade agreements with the USSR, the Eastern Bloc and China.

A blind man could see that the days of MPCA and Bookers were limited. On the one hand, the MPCA was powerless to defend itself. The stinging verbal attacks that Minister Carrington exchanged with Richard Ishmael grew more vitriolic. On December 7, 1973, the Guyana Graphic published a letter the Minister of Labor had sent to Harry Lall, the President of GAWU "in the hope of testing public reaction to the matter." In the letter Mr. Carrington said his Ministry appreciated the fact that the MPCA did not speak for a large percentage of the fieldworkers. He further noted that " . . . over the years, the Ministry of Labor has bent backwards to accommodate the rivalry between the MPCA and your union . . . ," however, the policy of de facto recognition was " . . . causing mental and physical strain on my officers at the expense of other industries and other workers . . . . We must find a way to bring an end to the conflict between your union, the GAWU and the MPCA . . . ." Obviously, the MPCA's usefulness in containing labor disruptions in the industry had ended and an increasing number within the ruling circle was prepared to dump the MPCA.

Ever vigilant, Bookers had been preparing for the eventuality of nationalization since the political crisis in the early 1950's. Immediately after the suspension of the constitution in 1953, Jock Campbell announced that any future investments in Guyana would have to be hedged by a policy of geographic and functional diversification. As Campbell explained it:

> It is, after all, a change in emphasis of policy, and in timing, which events have compelled, rather than a change in fundamental policy. For many years the Booker Group has been developing investments outside Booker Groups--in Central Africa and in the United Kingdom. Now we have been forced to accelerate this process in order to underwrite the political hazards of British Guiana. Your Boards are confident that you would not wish them, by applying this change of emphasis too rigidly, to lose profitable opportunities of purchase or partnership which suit and will strengthen the Bookers Groups.[46]

There were three outstanding features of Bookers' hedging policy over the next twenty-odd years. Firstly, Booker did not panic when confronted with general political instability or direct challenge. Therefore, in the midst of the chaos, triggered by the strike against the 1963 Labor Relations Bill, the Chairman reassured shareholders that " . . . Booker remains resolutely on course in the stormy waters. We shall not turn to other ports or scuttle the ship."[47] Instead, Bookers consciously sought to adapt itself to the rapidly changing political scene in the underdeveloped world. That was the whole point of Guyanization, a policy of indigenization which Bookers also used successfully in other Third World countries.

Secondly, Bookers' hedging policy was defensive and inconsistently pursued. When political tensions mounted, Bookers would quicken the pace of its acquisition in the

industrialized countries. Thus, during the struggle for
independence in Guyana, Booker steadily increased its
investments in food distribution and engineering in the
United Kingdom and merchandising in Canada. However, as
the political crisis receded, the pursuit of a geograph-
ical and product mix also lost its vigor. Bookers ran
into many problems during its first major attempt at
diversification. The investments in Canada proved a
complete failure and Bookers' United Kingdom shops and
engineering companies were only marginally profitable by
the end of the sixties. Nonetheless, a diversified
foundation had been laid; and, more importantly, Bookers
had acquired invaluable experience which would serve it
well during its next crucial phase of diversification
which began in 1970.

Thirdly, despite the political, economic and climatic
risks associated with tropical agriculture, Bookers
remained firmly committed to sugar. There were droughts
followed by floods, a boom-bust price cycle and, according
to Bookers, a five percent rate of return on capital
investments. Notwithstanding these adversities, Campbell
reported "Once again we have made the most of sugar in
desperately difficult conditions--and of rum: and they
have made the most for Bookers. But this cannot always
happen . . . I believe that sugar will remain the mainstay
of Bookers, there will be bad sugar years, even bad sugar
cycles."[48] In 1967 Bookers made more explicit this
presupposition governing its diversification program " . .
. it does not make sense--economically or socially--to
believe that any advantage would be gained by diversifying
at the expense of sugar."[49] In sum, Bookers was seeking
a balance between its investments in the underdeveloped
countries and the United Kingdom, because of political and
economical uncertainties in both parts of the world.
Bookers was not, we emphasize, looking to get out of its
traditional businesses or Guyana.[50]

Bookers relaxed its program of geographic diversification after the PPP was maneuvered out of office.
However, as the PNC's development strategy began to assume
an increasingly leftist orientation starting in 1969,
Bookers once again looked to the Motherland for security.
Bookers was particularly disturbed by (1) the nationalization of Alcan< (2) the creation of the External Trade
Bureau, which radically restructured Guyana's distributive
trade, (3) strict price controls on stockfeeds manufactured by Bookers, (4) the competition from the newly
formed Government printing company and (5) Burnham's
refusal to increase the local selling price of sugar.
These differences, along with the increasing divergence of

opinion with the Government over how to best handle the mounting industrial disruptions in the sugar industry betokened the end of an alliance which had united Bookers and the PNC since the late fifties.

Thus in 1970, Bookers embarked on a new policy; henceforth, it would attempt to earn at least 50% of its profits from its United Kingdom businesses. Moreover, Booker's new Chairman, George Bishop, who placed discernibly more emphasis on profits than had Campbell or Powell, reassured stockholders that Bookers had greatly increased the pace of acquisition of large wholesale, retail and self-service shops throughout the United Kingdom. Bookers also began a massive capital investment program to expand and modernize its engineering division. Notwithstanding these defensive strategies, Bookers was not ready "to scuttle the ship" in the Caribbean. Bishop's 1973 statement to shareholders is worthy of lengthy quotation. It underscores the essentially unchanged character of Bookers' hedging policy and its on-going commitment to sugar, shops and ships and Guyana:

> Our basic aim remains the same—to secure that half our profits come from the United Kingdom and from overseas. In these tempestuous days of rapid change in the world's political and economical scene none of us can hope to foresee what lies ahead. We feel it essential to aim at a balanced company with a solid base both at home and overseas so as better to withstand the hard knocks which are inevitable in times of change. In some years, we shall benefit from our earnings at home; in others it will be our overseas profits that will sustain us in meeting difficulties in the United Kingdom.
>
> In shaping Booker McConnell for the 1980s we are aiming to have as its core a broadly based food company . . . . A very large part of our existing business depends on agriculture, food, drink and shopkeeping . . . . Even our shipping and engineering division have a connection with food. Much of our shipping business is concerned with the movement of sugar and our most consistently profitable engineering company makes sugar processing machinery.
>
> . . . Our belief in sugar and our continued expansion, not only in Guyana but also with our new investments in Africa, will provide their reward in 1974 and later.[113]

Although George Bishop did not share Lord Campbell's legendary vision of corporate responsibility and adaptation, the new Chairman did recognize that continued success and survival required finding a new modus vivendi with Third World states increasingly exercising their sovereign rights.

Sugar Price Bonanza

The forecast for the free and protected sugar markets for 1972-74 were very promising. In 1971 the Chairman of Booker McConnell told shareholders that "we expect considerably higher revenues over the next three years," from sugar. Between 1961 and 1971 the average CSA price per ton had been Ь45 (pounds); in 1972 it rose to Ь57. The price of sugar sold in the protected U.S. market also increased from Ь65 to Ь69. Moreover, the increasing level of sugar consumption was putting serious pressure on current production capacity. In 1974, the World Food Conference predicted that an additional 30 million tons of sugar would be needed within the next decade to satisfy the rising world-wide demand. By mid-1974 Bookers' Chairman noted, with satisfaction, that "With the world economic situation tending to move in favor of the primary producer, the countries in which Booker McConnell operates overseas have good prospects of growth."[52] This was definitely not a time when Bookers wanted to be pushed out of Guyana. Despite the Governments new assault on the private sector and the anti-imperialist diatribes of the PNC, Bookers intensified its established policy "to work with the Government of the day." With the prospect of nationalization looming on the horizon, Bookers demonstrated its commitment to the PNC's goal to develop and diversify the economy. Bookers submitted a 10-Year Development Plan to dovetail with the Government's 1971-80 Plan which ironically failed to materialize. Bookers' Stores and Guyana Distilleries Ltd., became public companies when shares were issued locally in 1970 and 1971. A bus assembly plant and a photographic color processing plant were opened. In 1973 drug manufacturing was expanded to replace facilities closed in Jamaica and Trinidad; a new shrimp processing plant and a tea blending and packing plant were also opened. In 1975, Hi-Flex Guyana, an enterprise for local manufacturing of hydraulic hoses, was opened with much fanfare.

Bookers wanted the Government to fully appreciate that it was not only running the most efficient sugar industry in the Caribbean, but that it had also placed "at the disposal of the various administering agencies of the Government, those skills and facilities which the Group, as a consequences of its many-sided activities, has been able to establish and develop within the changing economic and social context of Guyana."[53]   The Project Evaluation Unit (PEU) created in 1973, was the most concrete expression of this long-term policy. Its purpose was to evaluate diversified investments opportunities in Guyana and to work closely with the Government on various experimental projects. "In the normal course of its varied activities, the Booker organization has participated or assisted in projects connected with the University of Guyana, The Pegasus Hotel, the World Bank, Livestock Development Project, the Sea Defense Program, the Agricultural Machinery Investigation and Development Center, the Resource Development Study . . . the new Shrimp Processing Plant, and Cotton Research."[54]   In cooperation with the Government's efforts to diversify agriculture, Bookers undertook experimental projects including wing bean, cassava, cotton, soya bean, cane trash feeds, fruit, honey, paper pulp, goats, trawlers and industrial alcohol. "Disease, or other adverse factors, not the inability or unwillingness to invest, have prevented their ultimate commercial development."[55]   Obviously, Bookers was not going to repeat the errors of inflexibility and arrogance which had precipitated the nationalization of the Canadian and American Bauxite Companies. Instead, the company re-emphasized its role as the "good corporate citizen" which Jock Campbell had initially chosen for Bookers during the early phase of decolonization. The fact that the value of Bookers investments in its nontraditional businesses in Guyana were small in comparison to those in sugar, shops and ships ($22.3 million vs. $192.8 million) did not detract from their public relations effectiveness.[56]   Top Government officials and the Government controlled press, pointed to Bookers as an example of the role the private sector should play in Guyana's economic development. Thus, despite the increasing pressure to nationalize the sugar industry within Caribbean intellectual and political circles, and the increasing radicalization of the PNC's development strategy, the Government was not eager to dissolve its partnership with Bookers. It took a catastrophic crisis, precipitated by extraordinary events in the international economic system to upset the balance of economic, political and trade union forces which had characterized the sugar industry since the PNC took office in 1964. By the start of 1974, the Guyanese economy had reached its nadir. Floods followed drought in 1973 and

sugar production fell to a 10 year low. Rice exports had
also suffered. The world economy was in a tail spin
created by monetary instability, inflation and an oil
crisis. Guyana ended the year with an $80 million balance
of payment deficit and its foreign reserves at an historic
low. Reserves continued to drop alarmingly during the
first weeks of the new year.

The price explosion which began 1973 in the world
sugar market, therefore occurred when all other hope had
faded. Sugar would have to earn the foreign exchange to
pay the inflated prices for imported fuel, fertilizers,
capital goods and consumer items. Sugar would also have
to supply the Government with revenue to support its
rapidly expanding bureaucracy, military apparatus and
ambitious development schemes. In sum, external developments had driven the PNC into an extreme dependence on
sugar. The Government had no alternative; it was determined to exploit the unprecedented conditions in the world
sugar market. Additional pressure was added by the
realization that the price bonanza was temporary. In
order to reap the maximum benefit, the Government had to
ensure first, that production was maintained at the
highest possible level; second, that sugar was sold in the
most lucrative markets; and third, that the windfall
profits would not be expatriated to the industry's
foreign owners or consumed by workers in the form of wage
increases or bonuses. The PPP/GAWU likewise realized
the sugar price bonanza provided a rare and fleeting
opportunity for it to force the PNC to accept a left-wing
coalition Government. The PNC Government was inescapably
on a collision course with Bookers and Jagan's sugar
workers.

The PNC's subsequent strategy in the sugar industry
is consistent with James Petras' analysis of the emergence
of a state capitalist model of accumulation.[57] According to Petras, the state vacillates between alliances
with the left (the workers) and the right (principally
foreign capital). In short, it plays one side off against
the other in the interests of state-controlled accumulation. The strategy is basically two-pronged. First,
foreign capitalists bear direct responsibility and blame
for exploiting labor and repressing its attempts to
organize. Secondly, the regime squeezes an increasing
portion of the foreign-owned surplus into the Government's
treasury. When workers demand a larger share of the
product of their labor, the Government appeals for peace
in the name of national interests or, even more explicitly, sides with capital, i.e., uses the armed forces to
break strikes. When, on the other hand, foreign capital

complains about the creeping confiscation of the company's profits, the Government reminds the private owners that the nations' resources belong to all of the people. In Guyana, the ruling party effectively used Bookers to deny sugar workers' demands for a decent standard of living and democratic trade unions rights for over a decade. Then, when world sugar prices exploded, the Government appealed to labor and dumped Bookers.

The world sugar market was in chaos. Negotiations to conclude a new International Sugar Agreement (ISA) had collapsed. Discussion concerning Britian's entry into the European Economic Community (EEC) and, consequently, the quota for Commonwealth Sugar Producers had also stalled. The United States had threatened to cut the Caribbean sugar quota, but two weeks later had increased it by 40,000 tons. Moreover, prices on the free market had wildly fluctuated between Ł93 and Ł274 per ton during 1973. Predictions were for even higher price jumps in 1974. Meanwhile, the CSA prices stood at a mere Ł61 per ton. Under the circumstances, Burnham and other Caricom Heads-of-Government, intervened in the marketing of sugar which, up to now, had been exclusively handled by the West Indian Sugar Association of which Bookers was a leading member. The world price stood at Ł200--to sell at the CSA price of Ł61 would be a give away. In January, Burnham ordered the suspension of all sugar shipments to the United Kingdom until a higher CSA price could be negotiated.[58] In February, Britain agreed to increase the CSA price by Ł22 to to Ł83.[59] The world price explosion continued. Production shortfalls were being announced everywhere and by August 1974, sugar sold for the unbelieveable price of Ł360 per ton on the world market. By November, the world price hit Ł650 per ton.

Burnham again suspended shipments to the United Kingdom. The outstanding portion of the Guyana quota (136,000 of 190,000 tons) would not be shipped until the CSA was substantially increased. Furthermore, this time, the traditional pattern of Caribbean leadership "kow-towing and salaaming, lobbying and pulling at jacket sleeves" in Whitehall was reversed. In September, the British Minister of Agriculture, Mr. Peart, called on Prime Minister Burnham. Britain's former colonial sub-jects were intoxicated by the whole affair. "We cannot but smile behind the hands at the spectacle of a Guyanese Prime Minister summoning a top ranking British Cabinet Minister to fly down to our country to discuss sugar prices . . . . The entire nation, irrespective of politi-cal viewpont is behind our Prime Minister in the

negotiations, so vital to the country's well being
perhaps survival may be a better word."[60] The negotiations led to a new CSA price of Ƚ140 per ton, an increase of Ƚ57. In addition, it was agreed that Britain
would buy only 85,000 tons instead of the promised
136,000. This left Guyana with an additional 51,000 tons
to sell on the world market at windfall profits.

While Guyanese were thrilled by the world price
bonanza, Bookers' response was one of concern and caution.
In the post-war period, the now increasingly threatened
CSA had provided the guarantee underlying Bookers' massive
expansion and modernization program. Sugar price booms
were always welcomed and greeted as the just reward for
confidence shown during the long lean years; but, BSE
preferred the assurance of stable prices and reliable
outlets. The chaotic world market and the impetuous
actions of the Guyanese Government now threatened all of
this. Although Bookers said it "recognized the need for
the Government of Guyana to oversee the marketing of sugar
. . . ,"[61] the doubts of one more experienced in such
matters were obvious:

> . . . it is not in the producers' long-term
> interest that the sugar agreement should have
> collapsed and that the price for marginal free
> supplies should have leapt to these levels. High
> prices defeat themselves. Developing countries
> need to secure sufficient guarantees of markets
> at remunerative prices against the day when the
> pendulum swings the other way--as it surely will
> . . . . These countries do not have the economi-
> cal resources to withstand a sustained period of
> low prices . . . . "[62]

It was not as if Burnham's Government was oblivious
to the day when the pendulum would swing the other way.
Throughout the sugar bonanza of 1974 and 1975, Burnham
pledged that Guyana would fulfill her CSA commitments;
however, since "charity begins at home" he also announced that Guyana would break with the established policy
of picking up the shortfalls of other Caribbean producers.[63] In response to Bookers' foreboding, Burnham
stressed the existence of new markets in non-Western
countries.[64] The table below shows the destination
of Guyana's sugar exports in 1974.

TABLE 1:   GUYANA'S SUGAR EXPORTS, 1974[65]

Destination	Tonnage
United Kingdom	127,244
United States	102,722
Morocco	16,090
China	10,396
Surinam	560
Venezuela	11,850
Finland	13,900
Tunisia	13,673
Algeria	1,760
TOTAL	301,165

During the fall of 1974, the General Manager of the Cuban Sugar Corporation visited Guyana's sugar industry. Talks with Government officials centered on Cuba's marketing strategy. In 1975, the Government's joint marketing committee with BSE found additional outlets in the Soviet Union, Denmark, Japan, West Germany, and Angola at vastly higher prices than those paid by Britain.[66] Guyana had clearly "cashed in on its non-aligned policy"[67] but a closer look at the table quickly demonstrates Guyana's continued and overwhelming dependence on traditional Western markets. The new markets simply offered Guyana a new, limited manoeuvreability so long as the world market remained bouyant.

Far more important than the Government's foray into sugar marketing was its decision to impose an export levy on the windfall profits in the sugar industry. In June 1974, with the world sugar price quickly heading for ₤ 300 per ton, the Minister of Finance, Mr. Hope, came before the PNC Parliament and asked that it enact the Government's bill imposing a levy on sugar exports retroactive to January 1974. Minister Hope's introduction of the bill is noteworthy for its presentation of the Government's rationale generally unencumbered by the usual public relations embellishments. For instance, there are no references to the socialist revolution, the teachings of Marx or Lenin, or even the right of developing nations to own and control their natural resources. Instead, Minister Hope embarked upon a lengthy description of the multifaceted economic crisis confronting Guyana. There was the international monetary crisis, devaluation, the rising cost of imports, high interest rates, an oil crisis, renewed rounds of financial instability, soaring inflation on industrial and oil based products, deteriorating terms of trade for UDCs now faced with the "real possibility of a collapse of their economies because of inability to met their external payments." In 1974, Guyana's import bill for petroleum products would be $70 million higher than in 1972 for the same volume. When price inflation for machinery, trucks, tractors, fertilizers and other indispensable imports were added, the cost increases were estimated to be as high as $150 million. "We face, therefore, this real problem of finding more than $150 million extra to keep the economy of Guyana going in 1974."[68]

In addition to this argument of necessity, Minister Hope pointed out that the PNC Government had played a vital role in negotiating the higher prices for sugar and locating more lucrative markets. Under the circumstances, " . . . the time is certainly right when we should, as a

Government, ensure that a fair proportion of the extra
earnings of the industry remains in this country for the
development of Guyana and to ensure that the balance of
payments of Guyana is favorably affected." In sum, this
was not an ideologically motivated attack upon the international capitalist system or imperialism. In fact,
Minister Hope stressed the fact that ". . . nothing that
we have done implies any objective of imposing any undue
burden on the industry. What we are taxing is the super
profits."[69]

As to Bookers, "what the Government would like to
see . . . is expansion and expansion."[70] Bookers
accepted the levy as reasonable and justified under the
circumstances.[71] Moreover, BSE continued with its
expansion plans. Over two thousand new acres were put
under cane in 1974. Bookers' motivation was the same as
always--profit. In 1974, BSE made a profit of $7.6
million despite regular taxes, an artificially low local
price for sugar and the new levy. Indeed, when the United
Kingdom-based companies did well just to hold their own,
the Booker Group in Guyana contributed 43% to the parent
company's profits.[72] Furthermore, the Government had
reassured Bookers privately and before Parliament that the
levy would be re-examined if it should be found to place
an undue hardship on the company.[73]

The sugar workers reacted with immediate hostility to
the levy. On April 26th, the Government first announced
its intention to impose a levy in a news article entitled
"New deal for sugar workers, says Hope, $30 million levy
to provide homes, bigger pensions." This would be done by
increasing the existing levy for the Sugar Industry Labor
Welfare fund (SILWF) from $4.80 to $12.00 per ton; the
Price Stabilization Fund (PSF) from $1.20 to $2.50 per ton
and the Sugar Industry Rehabilitation Fund (SIRF) from
$7.50 to $12.00 per ton. The SILWF and PSF would provide
the monies for the improved housing and pensions; the SIRF
would help Bookers finance the proposed expansion.[74]
The sugar workers were not impressed. They realized that
the new levy would make it more difficult to win wage
increase demands on the basis of the new higher prices
and, most importantly, that it would slash into the their
profit-sharing with the companies since the levy was
deduced before the calculation of profits. Under the
terms of an agreement with the SPA, sugar workers were
entitled to 60% profit-sharing on returns above 10% on
capital employed in the industry.[75]

The table below indicates the actual amounts sugar
workers received under the scheme between 1969 and 1976
and the amounts due if the levy had not reduced the
industry's profits after 1974.

TABLE 2:   PROFIT-SHARING PAYMENTS AND ENTITLEMENT

Year	Payments Made	Payments Due Sans Levy
1969	$1,469,310	NA
1970	NIL	NA
1971	196,246	NA
1972	867,122	NA
1973	NIL	NA
1974	7,956,568	60,906,209
1975	NIL	120,952,569
1976	NIL	41,475,000

(Source: Compiled from figure presented in a GAWU
Report entitled "The Case of the Sugar Workers.")

Meanwhile, the Government reaped handsome benefits from
the levy: $91 million in 1974, $227 million in 1975 and
$89 million in 1976. The indirect benefits promised to
sugar workers through increased Special Funds levy was a
mere pittance in comparison. The GAWU/PPP accuses the
Government of stealing more than $215 million from the
country's sugar workers.

The battle over the new industry surplus grew right
along with the price increases. As the Government negotiated with the British for higher CSA prices and searched
for more lucrative markets, the Government also indicated
that it would not tolerate the unions' attempts to flit
away the hard won increases in the form of higher wages.[76]
This was no hollow threat. In January 1974, the Minister
of Labor intensified his attacks upon trade unions operating in the sugar industry. After "bending backwards to
accommodate the union rivalry" his ministry was ready to
impose a Wages Council in the sugar industry and thereby
do away with trade unions altogether: " . . . it was not
our intention to destroy the collective bargaining process, but in an industry as vital as the Sugar Industry .
. . Government will have to consider doing something
. . . . "[77] The situation deteriorated rapidly following the Government's announcement of the new export levy.
Despite the fact that sugar workers had just returned to
work, the GAWU called another strike immediately. Outstanding demands were carried over but the "real cause" of
the present action was, according to a union official,
recognition and the levy which it estimated would rob
workers of over $18 million in profit-sharing.[78]

Events moved quickly throughout 1975. The GAWU
repeatedly disrupted production and reduced output by
40,000 tons precisely when contracts for future sales
reached their most profitable levels. Furthermore, when
the MPCA saw that the Government was preparing to betray
their long-term alliance, it began talking with the GAWU
about the possibility of a merger.[79] By March 1975, the
four major unions in the industry, the GAWU, the MPCA, the
NAACIE and Headmen's Union were united under the leadership of Dr. Cheddi Jagan to oppose management and the
Government.

Their demands, presented to the Ministry were as
follows:

(1) The levy must be replaced by an excess profit tax
     which would be deducted after the calculation of the
     company's profit-sharing with employees. The pro-
     ceed should be placed in the Price Stabilization
     Fund (PSF) and Labor Welfare Fund (LWF) in equal
     proportion.

(2) The money the Government had accrued from the levy
should be divided as follows:

   a) 50% to be paid to sugar workers, in proportion to
       their income;

   b) 25% to be placed in a Special Revenue Fund to
       provide cost-of-living allowances to <u>all</u> workers
       in the country whose earnings are less than
       $91.08 per week;

   c) 25% to be placed in the PSF to cushion any fall
       in future sugar prices and the LWF in equal
       proportion.[80]

Their demands went unheeded.   The government had already
spent the last year's levy revenue to finance its current
and capital accounts and planned to do the same in the
days ahead.

Burnham's Government immediately retaliated.   The
long-threatened compulsory arbitration legislation (Trades
Disputes Bill) was railroaded through Parliament on March
17th.   Why this amendment, which was so repugnant to
organized labor?

> While Nero fiddled, Rome burned . . . We have
> done too much over this period to steer this
> economy to make it viable, to come out on top of
> a most difficult situation to allow people who
> are confused to do this . . . Could we wait in
> every case for the trade unions to act if we are
> really interested in the workers?   They have been
> acting for over 25 years.   It is a big show, but
> the curtain must fall now.[81]

The existing labor law provided for arbitration when both
parties agreed; with this amendment, the Minister of Labor
was authorized to refer a dispute to arbitration, when
either party requested or without consent.   Moreover, the
Minister would choose the arbitrator(s) and decide which
issues in a dispute were to be considered by the tribunal.
A last minute but successful lobbying effort by the TUC,
eliminated a provision which would have empowered the
Minister to vary the terms, or duration, of the arbitration award at his discretion.   Notwithstanding this
concession to the TUC, the Ministry was given an enormous
power, especially when we consider that the Government
employed a rapidly increasing proportion of the population.   In essence, the Government, as the employer, could

now request the Government, as the Ministry of Labor, to
appoint an arbitrator and decide his or her terms of
reference and thereby settle the Government's differences
with its labor force.

The Government was girded for battle with the four
unions. It disregarded the warning from the United Force
that while the PNC could pass unpopular laws, it "cannot
throw all the workers who stay home in jail."[82] The
Government pressured Ishmael into breaking the MPCA's
brief alliance with the GAWU. The MPCA called off the
strike (a meaningless gesture) and agreed to binding
arbitration knowing that the companies had no money to
meet the union's demands because of the levy. The Crane
Arbitration Tribunal was set up by the Minister of Labor.
Intermittent strikes continued under the GAWU's leadership. In hopes of crushing the strike, the police, army
and national service were called in to intimidate strikers, protect scabs and harvest the cane.[83] In July, The
Crane Arbitration Award was announced: workers were to
receive wage increases of 20% for 1974 and 10% for 1975,
totaling $16 million for the two years.

The MPCA accepted, of course. The GAWU called the
award "an insult to workers" in light of the super profits
being made by the industry. A strike call again went
out. This time however there was an important omission
from the list of demands.

> The GAWU rejects the Crane Tribunal Award as
> unrealistic and unsatisfactory. It calls for a
> change of the award by the Government, the
> nationalization of Booker McConnell & Co., Ltd.,
> its recognition as the bargaining agent for field
> and factory workers, workers' control in
> decision-making and management, and adequate
> remunerations for the toilers who produce the
> sugar.
>
> It also calls on all sugar workers to wage a
> relentless struggle until their just demands are
> met--full employment in-corp and out-of-corp and
> at least the same wages of $11.68 per day,
> inadequate as they are, which their brothers in
> the bauxite industry receive.[84]

The GAWU was not, as before, demanding the payment of
profit-sharing for 1975 nor the replacement of the levy
with an excess profit tax. These were precisely the
points on which the Government would not and could not
negotiate--the money was already spent. Therefore, their
omission was clearly a signal to the ruling party that

Jagan was ready to bargain. With the country losing millions in foreign exchange and Government revenues, Burnham had no choice but to negotiate.

The ideological differences between the PNC and the PPP had been fading as Burnham's Government veered to the left during the early seventies. The United Force claimed that "People are saying that this Government's brand of socialism is about a hair's breath away from the Cheddi Jagan's brand of Communism."[85] Had not the PNC nationalized the Reynolds Bauxite Company in January 1975 and the Jessels Companies in May?[86] Were not PNC leaders and Government officials increasingly quoting Marx and Lenin as justification for their policies? Minister Carrington now chided sugar workers to be patient by reminding them that it had taken the Soviet Union (the PPP's definition of a good socialist society) fifty years to build socialism. Therefore, it came as no surprise when a PNC convention adopted a resolution in August 1975 which claimed that the party adhered to the philosophy of Marx, Engels and Lenin.

The PPP was pleased by the belated acknowledgement of the correctness of its line. "We, on this side of the House, gather from these happenings that the Government appears to be serious in moving towards socialism."[87] The PPP reversed its 1973 policy of "non-cooperation and civil resistance" and adopted a new line of "critical support."

> "Critical support" is basically intended to give a firm rebuff to imperialism, a warning that the PPP recognizing imperialism as enemy number one, will join forces with the PNC . . . "support" implies joining forces, with the PNC in creating national unity to protect Guyana's sovereignty and territorial integrity. "Critical" implies not only criticizing all short comings of the PNC and opposing all the sinister moves of the reactionary forces, but also making clear all the differences in policies, programs and ideology between the PPP and the PNC . . . . [88]

Publicly, the GAWU demanded recognition, nationalization, workers' management and higher pay for sugar workers. Behind closed doors, the PPP manoeuvred to gain a place in Government. According to Jagan, critical support was only the first step towards a political solution to Guyana's problems.[89] Its trump in these negotiations with the PNC was the unpaid profit-sharing due to the sugar workers controlled by the PPP. We lack direct evidence, but it

seems likely that Jagan implied that the PPP would forget
about the money once burdened with the responsibilities of
high Government office.

The PNC continued to resist the demands of the
GAWU/PPP but the advantage quickly slipped to its op-
ponent. The strikes had cost the Government $52 million
in lost revenue and $100 million in foreign exchange in
1975.[90] The price of sugar and bauxite was plunging on
the world market.[91] The Governor of the Central Bank
warned that if the strike continued, the Government's
development plans for next year would have to be "drastically revised."[92] Pressed to the wall, Burnham had no
alternative but to adopt the strategy championed by
Minister Carrington: recognize the GAWU and neutralize the
opposition in the sugar industry. The Government announced that a poll would be conducted to resolve the inter-
union representational dispute. On December 31, 1975
sugar workers were finally allowed to elect their own
bargaining agent. The GAWU won 98.7% of the vote.[93]

Therein, the PNC turned its back on the remaining
portion of the rightwing alliance which had brought the
party to power. For more than ten years, the Government
had closely coordinated its activities with those of the
SPA, the MPCA and the TUC in order to suppress industrial
disruption in the sugar industry. The decision to conduct
a recognition poll was made without even consulting with
its former allies. Ex post facto, the SPA, the MPCA and
the TUC were called in and informed of the new policy.
The MPCA refused to participate in the poll. The TUC
objected that it was a violation of established trade
union practice and the SPA initially refused to serve
notice of termination of their collective bargaining
agreement with the MPCA or to pledge cooperation with the
proposed poll.[94] Their protestations fell on deaf
ears. An economic crisis of external origin made it
impossible for the PNC to continue to resist the ideological and political pressures from its leftwing opposition.

Nationalization

A twenty year long partnership with Bookers was
reluctantly broken. On December 18th, the PNC laid the
legislative groundwork for the nationalization of the
Booker Empire. Article 8 of the Constitution guaranteeing
prompt and adequate compensation for lands compulsorily
acquired by the state was amended. Henceforth, (1) lands
would be assessed at their 1939 values; (2) the evaluation

so made would be final and without appeal to the courts;
(3) the Minister of Finance was empowered to pay for the
acquired property with Government bonds either in part
or in full.[95] Notwithstanding this preparation, the
Government did not take the initiative. It was Bookers
who first said, "Take sugar."

As Bookers' Chief Executive Officer pointed out, the
company and the Government had no important conflicting
goals in the industry. Until the last minute, they were
unanimous in the decision to keep the GAWU out. Furthermore, Burnham appreciated the fact that Bookers was
running the most efficient sugar industry in the region
and expanding production and profitability while other
Caribbean producers were declining. Bookers had also
willingly cooperated with the Government's efforts to
diversify the economy. Most surprisingly, Bookers had
accepted the Government's increasing infringement on its
corporate freedom of action with minimal objections. For
instance, Bookers simply warned of the need for secure
outlets and stable prices when the Government took over
marketing and price negotiations. Bookers was also
willing to work with the GAWU as the recognized bargaining
agent for sugar workers. Indeed, even after the Government's unexpected decision to recognize the GAWU, Bookers
figured it had at least another three years in the country.[96] It was   Burnham's refusal to renegotiate the
levy which proved to be the straw which broke the camel's
back.

Bookers was concerned about the levy from the start;
however, " . . . we were reassured only by our understanding that it was the Government's intention to regulate the
levy so as to remove the exceptional high element in sugar
prices, while allowing the sugar companies to make a
reasonable return."[97] In 1974, Bookers paid $91 million
to the Government in special levy, but BSE still made a
respectable profit of $7.6 million for the parent company.
However, the picture changed drastically in 1975. Input
costs rose rapidly and strike action reduced production by
40,000 tons from the previous year. This, plus the levy,
led BSE into a loss of $4.8 million. If the Government
would not make adjustments in the levy, it was projected
that BSE would lose $15.5 million in 1976.[98] Obviously,
this was unacceptable to Bookers.

In February 1976, Bookers' Chief Executive Officer
personally came to Georgetown to discuss the matter with
Burnham. Mr. Caine explained the bleak situation confronting BSE. Surely, the Government would reduce the
burden imposed by the levy under the circumstances. To
his surprise, Burnham refused. His Government had already

invested millions in various development projects that required additional monies in order to return any benefit to the country. Burnham asked Bookers to help him out just for another year or two. Caine explained that this would be impossible, his Board was not willing to finance BSE's losses internally. Instead, he offered Burnham a partnership in the company--49% interest with Bookers holding a management agent contract. Burnham refused. According to Caine, "He got us into such a position, we had to tell him to take sugar." "Increasing Government intervention . . . resulted in the commercial viability to BSE being taken out of our control."[99] Nonetheless, Bookers was eager to hold on to its other businesses in Guyana. But again Burnham was uncompromising. If he took sugar, he would also take the entire Booker Group of Companies.

Burnham's intransigence stemmed from three beliefs-- all of which ironically proved to be wrong. First, he thought that he would seal the cooperation of the GAWU/PPP with the nationalization of the sugar industry. All of the explicit demands of the GAWU/PPP would be satisfied. Moreover, Burnham figured that two parties, now so ideologically similar, would be able to cooperate in the industry. He was completely surprised when the PPP waged the most devastating strike in the history of the industry in 1977, supposedly over the withheld profit-sharing payments for 1974, 1975 amd 1976.[100] Second, Burnham did not believe the figures Bookers produced to support its case for a reduced levy. Reynolds had done the same to fight the bauxite levy and the Government proved them a liar after taking over the company. Naturally, Burnham believed that Bookers was also hiding large profits through intra-company transfers and bookkeeping tricks. Notwithstanding their experience with bauxite, the Government was wrong. In 1976 and 1977, the state-owned sugar company would also lose money. Third, despite the fact that prices were already falling, Burnham had no idea they would drop as low as they ultimately did.

Of course there were other considerations which had long been pushing towards nationalization. The majority of Caribbean Governments had already acquired part or complete ownership of their sugar industries. Burnham, with dreams of regional leadership, did not want to be the last country in the Caribbean with a foreign owned sugar industry. In addition, the PNC had promised the people ownership and control of their natural resources since 1970. The Booker empire was an embarrassing challenge to this promise. And, on a more practical level, the Government was confident that it could competently run the

industry.   Thanks to the Guyanization of the top professional and managerial levels of Bookers' operations,
there would not be a shortage of local and qualified
personnel.[101]

Therefore, if Bookers wasn't willing to help an old
friend out, the Government was ready to remove Bookers
from the picture altogether.   It had all been a matter of
timing anyway.   Since the early 1950s, Bookers had realized that the post-colonial Government would nationalize
its property.   Similarly, the PNC Government knew that it
would eventually have to take-over Bookers' holdings,
including the den of its political rival.   Nonetheless, it
was to their mutual advantage to work together as long as
possible.   In the end they parted as friends.   Burnham
announced the proposed nationalization without the usual
saber rattling: "We take this step of fully domesticating
this giant that has in the past dominated our political,
economical and social life without any bitterness or petty
spite, but as a matter of course.   This was inevitable."[102]

Bookers was prepared for the inevitable.   For the
past five years, management had worked feverishly to gird
the company for the loss of its Guyanese businesses, the
traditional backbone of the company.   The latest hedging
policy--rapid acquisitions and expensive capital improvements in the Engineering and United Kingdom Food Distribution Divisions--paid off handsomely just in the nick
of time.[103]

> It is a remarkable tribute to the resilience of
> the re-shaped Booker McConnell that . . . The
> profit including the Guyana Companies in 1975 was
> surpassed by the profit in 1976 without those
> companies.

Following some minor share issues, slightly increased
loans and two major acquisitions in 1976,

> . . . the overall financial position is stronger
> than it was at the end of 1975 . . . 1977 had
> begun well.   The Board believes that Booker
> McConnell should earn a substantial increase in
> profit this year.   Relieved of the risks of
> tropical agriculture and the other uncertainties
> of Guyana, the Board can concentrate on the
> growth of our existing businesses.[104]

Of course there were the ritual recriminations; but
more importantly, the Government and Bookers amiably
worked through the six weeks of negotiations with a common

view of their future.[105] The opposing teams of ne-
gotiators wrangled over the Government's insistence on an
assets only acquisition, Bookers' insistence that it be
paid for standing cane and other improvements, the price
the Government would pay, the interest terms and the
repayment schedule, and a host of minor issues.   In a
sense, the Government had Bookers over a barrel.   Bookers
no longer wanted sugar because of the Government's levy;
but, the Government was the only buyer in the market for
Guyana's sugar estates.   Near the end of the negotiations,
Mr. Caine, the head of the Bookers' negotiating team,
woefully complained " You have complimented us on our
tactics.   Yours have been dogmatic, if tough.   On the
other hand, you hold most, if not all, of the cards . . .
we must reluctantly accept . . . . "   The Government was
unshakeable in its stand.   One, that it would take only
assets.   Naturally, this left Bookers with liabilities and
the responsibility for winding up the old businesses.
Two, that it would pay for acquired assets only the
written down book value for tax purposes.   Nonetheless,
the negotiations were not the slaughter that they might
initially appear to be.   Mr. Caine concluded the above
statement by noting that " . . . at the end of the day, it
is not impossible that the other elements of the final
package would be such as we describe as reasonable."

It turns out that Bookers and the Government had
already agreed to a split decision.   The Government would
win the first round of negotiations concerning the terms
of compensation and Bookers would win the second round
concerning the Government's purchase of marketing, technical and consultancy services.   Bookers had decided that
its excellent reputation in Third World countries for
social responsiveness, training of local personnel,
cooperation with government's development plans, and
flexible management was more important then the uncertain
gains of a vicious battle over compensation.   Moreover,
purchase of services agreements held out the promise of
dependable earnings without the risks of capital investments.   Even before the actual negotiations started,
Bookers had offered and the Government had accepted
"commercial arm's length arrangements between Bookers
McConnell Companies in the United Kingdom and the Caribbean and the businesses in Guyana."   Negotiations to
work out the details of this on-going relationship began
the day after Bookers accepted the Government's tough
principles on compensation.   "The general tone of the
meeting was friendly and constructive," markedly different
from the recent price meetings.   According to Bookers'
Company Secretary who participated in the compensation

negotiations, the second round of meetings were a snap.
"Those guys would start at 11:00 and finish at 2:00.
They'd be laughing and joking as we made the most grudging
progress."

Notwithstanding the Guyanization of most technical
and managerial positions within the Booker Group of
Companies, Guyanese are not able to run their businesses
without outside assistance. This has nothing to do with
the competence or incompetence of local personnel.
Bookers' businesses evolved as subsidiaries of a vertically integrated multinational operation. In order to
remain profitable, the companies must maintain these
linkages at least until new reliable relationships can be
established. Guyanization had not prepared local personnel to handle international sugar marketing, price negotiations or overseas purchasing. Moreover, even in terms
of local operations, Burnham accused "Bookers of keeping
the key technical knowledge in the hands of expatriates to
ensure BSE's continued dependence." These are the harsh
realities which any Third World government trying to
negotiate a better position with the international
capitalist system will have to confront; and, which most
often reimpose dependency. Thus the Guyanese Government
had little choice but to conclude agreements with Bookers
for technical consultancy, sugar and rum marketing services and overseas purchasing. For a fixed term and set
fees, the Government obtains the expertise and contacts it
needs in the world market. Moreover, the contracts
contain provisions for the training of local personnel to
fulfill these functions in the future. Indeed, the new
relationship with Bookers must be minimally satisfactory
since the contracts were renewed by the Government in
1979.106

Hopes and excitement ran high around the time of
nationalization. At the opening session of the negotiations for acquisitions, the Minister of Agriculture
reminded the participants of the enormous significance of
the steps they were about to take,

> Today, we meet to write a brighter and long
> awaited chapter in the history of Guyana and of
> Sugar. This historic occasion represents the
> dawn of a new era which will secure for sugar
> workers their final emancipation, and for the
> nation of Guyana the attainment of its economic
> independence from alien occupiers of its land and
> owners of its industry.107

The socialist path to development now seemed to lie straight ahead. Bookers, the largest and the last surviving expatriate owner, was finally being pushed out. The recent steps towards a political rapprochment between the country's Blacks and Indians also betokened a promising future. For the first time in more than twenty years, Burnham and Jagan mounted the same speaker's platform at the May Day Rally. Patriotism was, they claimed, stronger than their long-term rivalry. Jagan agreed that the country faced a grave external threat. "There is no doubt," Burnham warned, "that there is a planned and cynical attempt to invade and/or destablize our country."[108] To prevent this usurpation of sovereignty, the PPP and the PNC promised to unite and to resist the forces of imperialism at all costs.

> If we must die, Oh let us nobly die,
> So that our precious blood may not be shed in vain.
> Long live the people's struggle!
> Long live the revolution!
> Long live the Co-operative Republic of Guyana!!![109]

Moreover, they pledged themselves to work together to reach their common goal of a socialist society. The vesting day was carefully chosen. It would be Guyana's tenth anniversary of independence. Thus on May 26, 1976, at 12:01 a.m. the Government took complete ownership and control of the Booker empire in Guyana. Despite heavy rainfall, more than 40,000 people, mostly indigent sugar workers, turned out for the joint celebrations. Minister Kennard had called this their "final emancipation" and according to news reports, the crowd was elated. Unfortunately, sugar workers soon discovered that the heavy rains were a more faithful portent of what lay ahead than were the promises made that day.

Throughout its long history in Guyana, the sugar industry has been characterized by extremely poor industrial relations. Now, the Government was seeking to purchase the elusive and highly prized industrial peace by granting the GAWU recognition and assuming state control of the industry. To its surprise, the Government has also failed. Notwithstanding the initial industrial peace, the Government had not addressed the deep rooted difference between its party and the opposition. In exchange for critical support and an end to the wave of strike action in the sugar belt, the PPP wanted a place in Government. Moreover, Jagan was ready to use his powerful

political-trade union complex in the sugar industry to
achieve this goal.   As a result, the sugar industry
continues to serve as a political battlefield and the
sugar workers continue to be used as pawns by the rival
political parties.

Indeed, the longest strike in the history of sugar in
Guyana was waged the year after nationalization.   The
strike began August 1977. The GAWU claimed the issue was
the withheld profit-sharing for 1974, 1975 and 1976; but,
undoubtedly Jagan was distressed by Burnham's continued
reluctance to offer the PPP a place in Government.   The
political nature of the strike was obvious.   "Observers
. . . have pointed out that the strike is really aimed at
embarrassing the government or challenging its authority
since the union claims further a government imposed export
levy on sugar was what denied the workers the amounts
outstanding in profit sharing."[110]   In fact, there is
evidence to suggest that the strike was timed to reinforce
the PPP's proposal for the creation of a National Patriotic Front Government--in essense, a coalition government
(with the PPP in charge) designed to steer the country out
of its current political and economical crisis.[111]

The PPP knew there was no money to pay the profit-
sharing.   The Government had already spent the $407
million it had collected from the levy and the Treasury
and foreign exchange reserves were again in serious
difficulties.   Nonetheless, Burnham gave assurances that
the matter would be looked into.   In the meantime, the PPP
skillfully pushed demands for profit-sharing and a "political solution."   For instance, during a debate over
worker participation which took place shortly before the
strike, Jagan warned the PNC that their differences would
have to " . . . be sorted out at a higher political level
because the crisis will not be solved either at the sugar
estates or around whatever fund committee you have; it
will eventually be solved by the two parties."[112]   For a
while, Burnham equivocated and described the PPP's proposal for "an understanding and coalition between leaders
as superficailly attractive."   However, by September,
Burnham had flatly rejected the PPP's offer and denounced
its policy of critical support and the strike as a mere
ruse to obtain a share of political power.

The strike dragged on for 136 days. GUYSUCO, the new
state-owned corporation, outdid the former expatriate
owners when it hired 6,000 scabs to harvest the cane.   As
to be expected, the Government called out the police and
army to protect the scabs, initimidate the strikers and to
help harvest the cane.   Burnham also played the "race
card" and succeeded in turning the majority of Black

workers against the sugar workers.[113]   The strike was
finally broken.   Sugar workers failed to win back any of
the $218 million in profit-sharing the GAWU claims was
stolen by the Government's levy.   However, the Government
seems to have won a pyrhic victory.   The negative effects
of the strike are still being felt.   The prolonged disruption of cultivation wreaked havoc with tight planting
and harvesting schedules.   Moreover, the fruitless confrontation further polarized already embittered labor-
management relations.   In fact, it would appear that the
withheld profit-sharing has replaced the demand for
recognition as the GAWU/PPP's political lightning rod.
Whenever the leadership needs a popular issue to bring the
workers out, it has simply to renew its demand for the
$8,800 due each worker.   As recently as August 1979, the
GAWU raised the profit-sharing issue.

NOTES

1. Jay Mandle, "Continuity and Change in Guyanese Underdevelopment," Monthly Review, 28 (September 1976): 37-50.

2. Quoted in The Sunday Chronicle, 14 October 1973.

3. Booker McConnell, Ltd., Annual Report and Accounts 1964, Statement by the Chairman.

4. Scott, p. 194.

5. Ibid., pp. 199.

6. Parliament House, Debates, Georgetown, Guyana, 5 January 1970. (Typewritten.)

7. Ibid.

8. Scott, p. 183.

9. Ibid., p. 164.

10. The PPP charged that the Government was "raiding the fund" and that the industry had invested more than $5 million from the SILWF in Government securities. "The Labor Welfare Fund has not got the money because the Government is using it . . . we want to know what is the amount . . . where is it invested, and if they are invested in Government securities, the Government must unfreeze them so that the peoples' welfare can also be met." The PNC acknowledged the investment by the fund in Government securities but reassured the opposition that the Sugar Producer's Association had agreed to purchase the securities if the fund should be in need of cash. Obviously, Bookers was struggling to accommodate the Government. "Debates," 24 November 1972.

11. Bookers News, 13 February 1974.

12. Scott, p. 168.

13. According to Edgar Readwin, then Chairman of Bookers' Sugar Estates, Campbell offered Jagan's Government the opportunity to purchase 49% of the company's Guyanese businesses in the late fifties. Harold Davis, then a Booker manager, recalls that an offer for 33% was made in 1960. Readwin and Davis agree that Jagan rejected the offer of joint-venture because he feared his constituency would regard such a move as a sell-out. According to Brian Scott who conducted extensive interviews with Lord Campbell, Burnham's Government was also offered a joint-venture. Negotiations fell apart when Burnham insisted that the shares be given to the Government. The details of these offers are hazy--Jagan denies

that such an offer was ever made and no records of the offers could be found at the London headquarters of Booker McConnell. Nevertheless it does appear that Bookers was read to participate in a joint-venture with the government holding up to 49 percent so long as Booker retained management control.

14. Booker McConnell, Ltd., Annual Report and Accounts 1964, Statement by the Chairman.

15. The Sunday Graphic (Georgetown), 18 October 1970.

16. Interview with Anthony Haynes, Chief Executive Officer Booker McConnell, Ltd., London, England, August, 1980.

17. Booker McConnell, Ltd., Annual Report and Accounts 1972, Statement by the Chairman.

18. Bookers News, 3 September 1975.

19. Interview with John Huddart, former Personnel Director of Bookers' Non-Sugar Estate Businesses, London England, August 1980.

20. Norman Semple, "Industrial Relations in Guyana," paper presented at the Caribbean Regional Seminar on Labor Relation, Trinidad, 1973, p. 12.

21. The Graphic (Georgetown), 9 September 1970.

22. The Graphic, 16 September 1970. This was of course true so long as the Government retained the legal framework of industrial relations inherited from the colonial state.

23. Interview with Mr. Haynes, ibid.

24. The Graphic, 27 Janaury 1971; the Chronicle (Georgetown), 14 February 1971.

25. Interview with Norman Semple, Chief Labor Officer, the Minister of Labor, Georgetown, Guyana, November 1980.

26. According to Cleveland Charran, the current General Secretary of the MPCA, Minister Carrington was motivated by a personal vendatta against Richard Ishmael, then the head of the MPCA and the Trade Union Congress. Jealous of Ishmael's power Carrington sought to strip the MPCA of its certification and membership. Interview with Mr. Charran conducted at the office of the MPCA, Georgetown, Guyana, October 1980.

27. The Graphic, 9 December 1973.

28. Ibid.; the Argosy (Georgetown), 10 February 1974.

29. The Argosy, 13 September 1970.

30. The Graphic, 25 September 1970.

31. For instance, whenever the GAWU's struggle with the SPA and the Ministry of Labor would reach a critical juncture, the PPP would take control over the campaign. "After seeing that the Minister of Labor was succeeding to a point with Mr. Lall in bringing about a return to work, Jagan decided to intervene and give instructions as to what should be done and to what should not be done." The *Argosy*, 13 September 1970.

32. The *Argosy*, 13 September 1970.

33. Between 1965 and 1975 the Government appointed six such bodies. Interview with Mr. Haynes.

34. The *Graphic*, 18 September 1970.

35. The *Chronicle*, 13 February 1971.

36. The *Graphic*, 14 March 1971.

37. The *Graphic*, 10 March 1971.

38. The *Graphic*, 18 February 1971.

39. The *Graphic*, 6 December 1973.

40. The *Chronicle*, 14 February 1971.

41. The *Graphic*, 28 February 1971.

42. The *Graphic*, 31 January 1971.

43. The *The Sunday Chronicle*, 20 June 1971.

44. Semple, p. 12.

45. The *Graphic*, 21 March 1971.

46. Booker McConnell, Ltd., Annual Report and Accounts 1954, Statement by the Chairman.

47. Booker McConnell, Ltd., Annual Report and Accounts 1962, Statement by the Chairman.

48. Booker McConnell, Ltd., Annual Report and Accounts 1963, Statement by the Chairman. During 1963 Guyana suffered a political crisis and a prolonged drought; however, prices on the world sugar market were described as "sky-high."

49. Booker McConnell, Ltd., Annual Report and Accounts 1967, Statement by the Chairman.

50. This remained Bookers' policy up to the moment of nationalization. This statement by Bookers' Chairman in mid-1975 reflects this strategy of balanced operations: "While we had serious problems in trying to adjust our businesses in the UK to the twists and turns of Government policy we were able to rely on the improvement in economic growth in the overseas countries in which we work." Booker McConnell, Ltd., Annual Report and Accounts 1974, Statement by the Chairman.

51. Booker McConnell, Ltd., Annual Report and Accounts 1973, Statement by the Chairman.

52. Ibid.

53.   *Bookers' News*, 1 August 1973.

54.   *Bookers' News*, 19 March 1975.

55.   Bookers' submission to the Government at the nationalization negotiations. Files of Booker McConnell, Ltd., London Office.

56.   Figures taken from Bookers' submission to the Government.

57.   James Petras, "State Capitalism and the Third World," in *Critical Perspectives on Imperialism and Social Class in the Third World* (New York: Monthy Review, 1978), pp. 84–102.

58.   The *Graphic*, 9 February 1974.

59.   The *Chronicle*, 23 February 1974.

60.   The *Sunday Graphic*, 15 September 1974.

61.   Booker McConnell, Ltd., Annual Report and Accounts 1974, Statement by the Chairman.

62.   Booker McConnell, Ltd., Annual Report and Accounts 1973, Statement by the Chairman.

63.   Ibid., The *Chronicle*, 23 February 1974. On the other hand, Bookers had always considered it a privilege to pick up the shortfalls of other Caribbean producers. Thus when CSA quotas were renegoiated, Guyana could argue for a larger share of the protected market.

64.   The *Chronicle*, 8 September 1974.

65.   The *Chronicle*, 9 March 1975.

66.   /The *Chronicle*, 20 November 1977.

67.   The *Chronicle*, 3 September 1975.

68.   Parliament House, "Debates," 19 December 1974. (Typewritten).

69.   Parliament House, "Debates," 28 June 1974. (Typewritten.) The Sugar Levy Act of 1974 provides for a levy on sugar exports at the following rates: (1) 55% on proceeds in excess of $365.00 per ton; (2) 70% on proceeds in excess of $521.00 per ton; (3) 85% on proceeds in excess of $625.00 per ton. Minister Hope was quick to point out that sales at the CSA price of Ł83 (pounds) would only be taxed at a rate of 4%.

70.   Parliament House, "Debates," 28 June 1974.

71.   Interview with Michael Caine, Chairman, Booker McConnell, Ltd., London, England, August 1980.

72.   Booker McConnell, Ltd., Annual Report and Accounts 1974, Statement by the Chairman.

73. Interview with M. Caine; Parliament House, "Debates," 28 June 1974.
74. The Graphic, 26 April 1974.
75. Booker McConnell, Ltd., Annual Report and Accounts 1968, Statement by the Chairman.
76. The Chronicle, 23 February 1974.
77. The Graphic, 11 January 1974.
78. The Graphic, 4 May 1974.
79. Parliament House, "Debates," 17 March 1975. (Typewritten.); The Chronicle, 17 November 1974.
80. The GAWU, "The Case of the Sugar Workers," n.p., n.d., available at Freedom House, PPP headquarters, Georgetown, Guyana.
81. Parliament House, "Debates," 17 March 1975. (Typewritten.)
82. Ibid.
83. Caribbean Contact, August 1975; the Chronicle, 1 May 1975.
84. The GAWU, Press Release, July 1975, available at Freedom House.
85. Parliament House, "Debates," 23 May 1975. (Typewritten.)
86. The Jessel Group of Companies which owned 20% of the country's sugar industry and an assortment of other businesses were nationalized in May 1975. Oliver Jessels, a British "financial wizard," bought the companies in 1971 and embarked on a bold scheme to transform the companies. The problems started almost immediately. He complained of heavy taxes and insufficient incentives; hundreds of workers were fired supposedly to improve efficiency. The Insurance Company went bankrupt; and finally, Jessels defaulted on the interest payments to its creditors. Jessels disposed of his easily realizable assets and indicated that he'd like to be rid of his Guyanese holdings. Amid scandal and rumor, the Government agreed to purchase Jessels' holding in Guyana for $15 million. One million was paid in the vesting day and the balance was due within 10 years at 8.5 percent interest. The Chronicle, 9 March 1975; Parliament House, "Debates," 23 May 1975.
87. Parliament House, "Debates," 23 June 1976. (Typewritten.)
88. Editorial, Thunder 7 (September–December 1975): 1–2.

89.   Ibid.
90.   The Graphic, 2 September 1975 and 17 November 1975.
91.   The Budget Speech 1976.
92.   The Graphic, 28 November 1975.
93.   The Chronicle, 7 January 1976.
94.   The Graphic, 30 November 1975.
95.   Parliament House, "Debates," 18 December 1975.
96.   Interviews with Harold Davis, Chairman, Guyana
        Sugar Corporation, November 1980, and Anthony
        Haynes, Chief Executive Officer, Booker McConnell,
        August 1980.
97.   Bookers' submission to the Government at the
        nationalization negotiations. Files of Booker
        McConnell, Ltd., London, England.
98.   Ibid.
99.   Booker McConnell, Ltd., Annual Report and Accounts
        1975, Statement by the Chairman.
100.   Interviews with Harold Davis, Chairman GUYSUCO and
        Norman Semple, Chief Labor Officer, Ministry of
        Labor, Georgetown, Guyana, October 1980.
101.   According to S. Bitar, MNCs must anticipate inten-
        sified pressures for local ownership and control as
        nationals increasingly master the technical,
        managerial and marketing skills involved in the
        operation of an industry. MNCs will, he advises,
        only remain welcome to the extent that they make
        available new technology and markets which the host
        country will in turn appropriate. Thus Bitar sees
        a recurring and mutually beneficial cycle of
        cooperation and conflict between host governments
        and MNCs which will lead to the development of
        increasingly advanced industrial sectors in host
        countries. The first part of Bitar's prediction is
        supported by events in Guyana. Unfortunately, the
        continuation of foreign private investment in
        increasingly advanced industrial sectors has
        not materialized. Bookers has neither sought to
        bring new technology or capital into Guyana or the
        Third World in general for that matter. S. Bitar,
        "The Multinational Corporation and the Relations
        Between Latin America and the U.S.," in Conference
        on the Western Hemisphere, Issues for the 1970s,
        Held at the Center for Inter-American Relations, 29
        April - 2 May 1971, pp. 85-104.
102.   Prime Minister Forbes Burnham, "Republic Day Speech,"
        23 February 1976.

103.   For instance, between 1970 and 1975 Bookers spent
       £5 million (pounds) to improve and expand its
       engineering division. The division made its first
       significant profit contribution in 1975, £1.86
       million versus £182,000 (pounds) in 1974. Booker
       McConnell, Ltd., Annual Report and Accounts 1975.

104.   Booker McConnell, Ltd., Annual Report and Accounts
       1976, Statement by the Chairman.

105.   The researcher was most fortunate to be granted
       access to Bookers' files on the nationalization
       proceedings which contained the submissions made by
       the company and the Government and the minutes of
       the meetings. The quotations and figures cited
       below are from this source.

106.   Booker McConnell, Ltd., Annual Report and Accounts
       1975.

107.   Gavin Kennard, Minister of Agriculture, opening
       remarks, Bookers' files of nationalization negotia-
       tions.

108.   Prime Minister Forbes Burnham, Radio broadcast, 26
       May 1976.

109.   Ibid.

110.   The Chronicle 1 September 1977.

111.   "P.P.P. Calls for a National Patriotic Front,"
       Caribbean Dialogue 4 (June 1978): 36-39.

112.   Parliament House, "Debates," 19 May 1977. (Typewrit-
       ten.)

113.   Caribbean Dialogue 4 (June 1978): 39.

CHAPTER 6

Conclusions

The transformation of the development strategy of the postcolonial state in Guyana unfortunately conforms to the grim predictions reached by Ian Roxborough, James Petras, Clive Thomas and Harry Magdoff in regard to the recent trend of radicalization in the Third World. Contrary to the official declarations proclaiming the end of class struggle and the great progress of the socialist revolution in Guyana, all of our evidence indicates that the Government's strategy of non-capitalist development has led to new forms of foreign dependency, economic stagnation and the increased repression of the working class. According to Roxborough, Petras, Thomas and Magdoff, this degeneration into neo-dependency is inevitable whenever the radicalization emerges as a bureaucratic response to the crises of underdevelopment and remains the project of the state sector elite acting as an independent class. The petty bourgeois state cannot sustain a confrontation with the imperialist system on its own. Notwithstanding the vitriolic rhetoric, the conversion to Marxism-Leninism and the adoption of socialist forms, such a narrow stratum quickly discovers that it cannot endure without either a mass base at home or friends abroad. The teetering state must either lean to the left or the right and for historical and current reasons the center of gravity favors the latter. Thus Guyana, like so many UDCs, has ended up with a Third World variant of state capitalism.

We firmly agree with Roxborough that it would be a practical and theoretical mistake to equate this foiled attempt at bourgeois revolution with socialist transformation or even with what some theorists of noncapitalist development refer to as "the transition to the transition." Petras' critique of these defiant regimes as pro-imperialist and Roxborough's characterization of their policies as lumpen-socialism are harsh but fundamentally correct.

The Devolution into Neo-Dependency

The international market conditions which led the PNC to challenge the inequity of the world capitalist system were very short-lived. Commodity prices which peaked in the early months of 1975 fell precipitously throughout the year and 1976. Inflation continued its sharp rise; major foreign exchange currencies continued to float widely and money remained tight and expensive on Euro-dollar markets. The talks to establish a NIEO stalled at an early stage. In sum, the economic forecast for underdeveloped countries

was once again gloomy. For Guyana the heady excitement of
the past eighteen months was over and the Minister of
Finance soberly projected a $214 million budget deficit in
1976. For 1976 the Government hoped to maintain the high
level of public investment and consumer subsidies by
financing the budget deficit out of the Government surplus
built up during the period of high export prices. However, danger was imminent and the Minister warned, "These
trends--high development expenditure, rising import prices
and not too significant gains in export prices--if they
materialize could seriously affect or even reverse our
balance of payment position, and swing the terms of trade
against this country."[1] The commodity price boom was an
anomaly. Nationalization of eighty per cent of the
economy and the forced repatriation of all foreign invested pension funds had not solved the problems of
structural underdevelopment. Development funds remained
in critically short supply. The foreign exchange crisis
reasserted itself with a vengeance. Reserves fell from
over $200 million at the start of 1976 to near zero at the
beginning of 1977. The Government was forced to severely
restrict imports and to cut the 1977 budget by 25 per
cent. Public sector investments were curtailed by 40 per
cent. The Gross National Product declined by 5 per cent
in 1977.

In January 1977 the PNC was forced to embark on a
desperate and humiliating search for foreign assistance.
First the Government turned to the Communist bloc. Both
the amount and speed of assistance were inadequate. Thus
no sooner than the Democrats regained the White House,
Burnham sent diplomats to Washington to repair the friendly and dependent relationship which had been severed
during eight years of Republican Administration. President Carter was eager to counteract Castro's greatly
increased influence in the region. Guyana went bankrupt
and for sale. A Guyanese official reportedly entreated an
American diplomat to "Please do what you can to help us.
It looks like there's nobody else."[2] The flood gates of
Western aid opened. In August 1977 the U.S. Agency for
International Development announced it was going to
increase its economic co-operation with Guyana by more
than tenfold.[3] In October Canada agreed to substantially increase its assistance over the next five years.[4]
In November the British Government signed a cooperation
agreement in which it agreed to provide G$50 million at 3
per cent repayable over 25 years for infrastructural
projects.[5] In December the Inter-American Development
Bank approved a loan for US$49.5 million to help finance
the construction of flood control, irrigation and drainage

works.[6]   Guyana was once again a favored client state.
In 1980 the American Ambassador to Guyana announced that
the country received the highest per capita aid from the
U.S. of all countries in the world.[7]   IMF lending is now
used to fill the balance of payments gap.   The economy is
trapped in an expanding web of debt and aid dependency.
In 1976, 7.6 per cent of the country's export earnings
were used to service its foreign debt.   By 1983 that
figure had risen to an astonishing 50 per cent.   Today
Guyana's foreign debt is estimated at US$1 billion.[8]

Since 1977 Guyana's economic policy has steadily retreated from the left.   Moreover, it must be recalled
that the Government never ceased to solicit private
foreign and local investments.   The Small Industries
Corporation was created to offer financial, managerial and
technical assistance to stimulate local investors.   A 1975
press release from the Ministry of Information and Culture
highlights the point that the Government continued the
search for foreign capital even while converting to
marxism-leninism.

> Apart from a stable economic and political
> climate, incentives range from a modern banking
> system to tax holidays and allowances, carryover
> of losses, duty free and other concessions.
> Markets range from those in neighboring Latin
> America and Caricom territories, to Socialist,
> Third World, and western hemisphere countries.
> It is the Government's firm decision that the
> investor is guaranteed the right to a fair and
> equitable return on his investment of capital and
> technological inputs and against any possibility
> of seizure, confiscation or expropriation of his
> investment.[9]

Therefore a return to a more traditional capitalist
development strategy did not go against the essential
grain of the PNC Government.   This was made clear in an
article written by a party ideologue in 1977.

> The dialectics of this period require a
> temporary shift to selected capitalist strategies
> for development.   The state sector is not an
> efficient producer of wealth and is likely to
> continue to experience complex transitional
> problems.
> Foreign private capital should be encouraged
> to invest with incentives that are even more
> favorable than similar developing countries.

> The local capitalist and the petty-bourgeois
> class should also be provided with opportunities
> to enhance the development process.[10]

Still the private sector demanded more specific assurances. By 1979 the Government responded by issuing its
long delayed "New Investment Code". It was readily
admitted that ". . . the economy would have to grow at a
more rapid rate than it has grown in the past. The
Central Government or the public sector alone cannot
generate all the resources necessary to ensure the required rate of growth."[11] Once again the Government was
publicly courting private capital as Sir Arthur Lewis had
argued would inevitably be the case. The Minister of
Economic Development made a great effort to portray the
new courtship as fundamentally different from the earlier
relationship but it was nonetheless clear that the Government had retreated from the militancy of the mid-
seventies. For instance, the Government retreated from
the policy of demanding 51% equity and effective control
and said instead that ". . . the level of equity participation will be determined on the basis of negotiation and
mutual agreement prior to establishment of the enterprise." Indeed foreign private capital could once again
invest without any constraint of joint-venture as previously required by the Declaration of Sophia. Even
strategic activities supposedly restricted to public or
co-operative development were so vaguely defined as to
allow private foreign investment in these preserves. As
the Code notes, "The very definition of the concept
'Strategic Activities' implies that it is not possible to
prepare a listing, invariant for all time. . . ."[12] The
confrontation with capital, particularly international
capital, was over. After a short-lived attempt to
transform international trade relations and to break the
stranglehold which multinational corporations had on the
economy, the PNC reaffirmed its dependency on international capital, technology, markets and aid.

The Crisis of State Capitalist Control of the Sugar Industry

The contradictions of state capitalism masquerading
as socialism are vivid in the sugar industry. The Government claims that in a socialist society, industry belongs
to the people and therefore workers should drop their
demands for profit-sharing and cooperate to increase the
level of productivity. Not surprisingly, sugar workers
scoff at these arguments. Instead, they ingeniously seek

ways to exploit the transitional problems of GUYSUCO.[13]
Moreover, the three major unions in the sugar industry
have rejected the Government sponsored plan for worker
participation in management. Critics charge that the plan
is a sham. At the time of our visit the GUYSUCO model had
yet to be operationalized so it is impossible for us to
make a firm judgment. On paper, it looks quite impressive. The plan calls for a four-tier management hierarchy
with substantial worker representation at each level.[14]
However, there is an apparent contradiction to the idea of
workers' democracy which does stand out. Final authority
to determine the agendas for the committees' meetings is
given to a member of management. If abused, this authority would make worker participation a meaningless exercise. Unfortunately, there is abundant, indirect evidence
that this is indeed the intention. First, this plan is
the brainchild of top management. Secondly, despite the
rejection of the plan by the three main unions, GUYSUCO
is, according to Mr. Davis, " . . . going ahead and
setting up the committees." In regard to the Special
Funds and Sugar Industry Labor Welfare Fund, it is
absolutely clear that the changes in the governance
structure were so designed as to keep tight control in the
hands of the Government. The Special Funds are now
controlled by a ten member committee. GUYSUCO has
four representatives, the Government has two, the workers
have three and the cane farmers have one. However, as the
Opposition was quick to point out " . . . we all know that
the Corporation is the Government even though they say it
is the people of Guyana. We know the people of Guyana do
not run it. So if the Corporation and the Government are
the same person with different clothing, where is the
justification for representatives of the Corporation,
representatives of the Government to drown the voice of
the worker."[15] Furthermore, the representative of the
cane farmers must serve on the National Cane Farmers'
Committee, the members of which are handpicked by the
Government. Thus, in reality, the Government controls
seven out of ten votes. In response to the Opposition's
charges, Burnham glibly remarked that even Lenin had
discovered that worker control takes time. Furthermore,
Government officers were, he claimed, just workers after-
all. The same situation prevails on the SILWF committee
which controls money traditionally used to develop housing
and community facilities for sugar workers. In addition,
the Government has now extended the definition of sugar
workers to include distillery, wharf and terminal employees, a large percentage of whom are Black and supporters on the PNC.[16]

In addition to these serious political and trade
union problems, the nationalized industry has also been
besieged by a depressed world sugar market and natural
disaster. World sugar prices dropped rapidly throughout
1976 and reached an abysmal low in 1977. Only two years
before, sugar had sold for Ᵽ650 on the free market, but
by September 1977 it brought a mere Ᵽ85 per ton--way
below the cost of production. The new owner's problems
were compounded by a simultaneous drop in output. In
1976, the industry produced 332,457 tons of sugar. In
1977, the year of the strike, production fell to 241,527
tons. In 1978, output rose to 324,805 tons, but it
nonetheless failed to reach he target of 360,000 tons. To
make matters worse, these short-falls were developing at a
time when the Government was under considerable pressure
to maintain the industry's reputation as a reliable
supplier to the now coveted protected markets in the
United Kingdom and the United States.

In 1978, smut disease attacked the cane fields. In
1979, rust disease was discovered. The diseases have
proven most hazardous to the normally most high-yielding
varieties of cane. Furthermore, GUYSUCO's technical staff
and Bookers consultants have not been able to control the
spread of either disease. Taken together the problems
have critically depressed the level of technical efficiency and output in the sugar industry. Factories are forced
to grind longer without the necessary level of maintenance
in order to produce smaller amounts of sugar.[17] The
crucial tons of cane, to tons of sugar ratio which Bookers' post-war drive for efficency drove down is rising
sharply.[18]

Profits depend on price, output and technical efficiency. Consequently, GUYSUCO is in serious trouble.
After the first full year of Government control, GUYSUCO
reported a $15.6 million deficit. In 1978, the situation
improved somewhat, but Mr. Davis warned that the near
future looked rather gloomy. After tax corporate income
was only $3.5 million and the rate of capital return was a
mere 3.8%. Furthermore, the company's liquidity continued
to deteriorate as income lagged far behind borrowing and
capital expenditures. The performance in 1979 was worse
than that of 1978; after-tax corporate income dropped to
$1.1 million and the rate of capital return fell to 1.7%.
The situation is clearly untenable. The Corporation
obviously cannot go on making huge capital expenditures
($50 million between 1977 and 1979) which bring only
marginal returns. As of 1979, the company was indebted
for $89.7 million, roughly the total value of its assets.

Mr. Davis tries to sound an optimistic note in his official reports and personal remarks; and, sugar prices were headed up again in the early 1980's after the severe depression of the late 1970s. However, it is unlikely that GUYSUCO will be able to produce enough sugar to satisfy its special marketing agreements and then have enough left over to benefit from the improved world market situation.

Mounting Labor Conflicts and Repression

The sugar industry is not the only site of labor-management problems. The bitter and frequently violent conflict between capital and labor also continues in other state-owned enterprises despite the policy of partnership. In fact, the struggle has worsened. In recent years, labor and political activists associated with the WPA have been murdered by law enforcement personnel and the rapidly increasing number of "political thugs" hired by the ruling party. A major confrontation occurred in the summer of 1979. Bauxite workers along with sugar and clerical workers mounted a national strike to demand the payment of a wage increase agreed to by the TUC and the Government. The Government, in violation of its own principle of binding consensus, had reneged on the agreement. It was also clear, however, that the OWPL, the GAWU, the NAACIE and the CCWU, the four major unions in the country, were hoping to destabilize the economy and force the PNC to resign. In a by now familiar scenario, the PNC charged that the strike was political and unleashed the coercive forces of the state. Gordon Todd, head of the clerical workers (CCWU), mysteriously disappeared. It was later learned that he had been arrested and beaten while in custody. Striking workers were attacked by club wielding law enforcement men who conveniently had removed their identification numbers. Eighty-two members of the CCWU were sacked and scab labor was hired. The bauxite workers were starved into submission after Government forces seized donated food supplies. Ultimately, the general strike was crushed. The fact that the primary goals of the unions were most likely political, does not, in our opinion, invalidate the labor's demands and grievances, i.e., wage increases, democratic leadership, etc. Nor, moreover, does it justify the brutal repression exercised by the Government. To call a strike political in a situation wherein the state controls over 80% of the economy is tautological and diversionary. The issues which the PNC hopes to avoid by this device is the legitimate civil and economic rights of workers.

The Government's anti-strike policy has even caused its collaborators embarrassment. Joseph Pollydore, General Secretary of the TUC, was moved to make the following critical remarks in regard to the regime's strike-breaking activities.

> We must seek to remove the situation which makes it necessary or expedient to resort to what is normally regarded as anti-socialist measures, such as, the recruitment of Army, National Service and other volunteers as recently happened to break strikes.[19]

Eric Huntley, a critic of the PNC, goes further and notes that the Government has made a habit of using Black scab labor in the sugar belt and thus continually fueled the country's racial hostilities. "It exposed just how low the government was prepared to go. In other words, they were prepared to use Afro-Guyanese against Indo-Guyanese so long as their own policies were promoted."[20] Notwithstanding the above, Mr. Pollydore has been careful to keep his criticisms within safe limits, and, more importantly, has refused to take action to force the Government to respect the rights of workers. In Pollydore's opinion,

> What the TUC does not, or what it refuses to accept is that it should join forces with others to take action against the Government as a means of preventing or overcoming these violations. The TUC prefers to pursue these matters by way of dialogue once the communication lines are open to it. . .[21]

According to Harold Lutchman, Pollydore's timidity is characteristic of the official trade union leadership. Despite the fact that "the policy of dialogue has been singularly void of meaningful results'," and the Government's persistence in making crucial national policy decisions without consulting the TUC,

> As matters now stand, many trade unions see themselves under an obligation to follow the line laid down by politicians or to do nothing to "rock the boat" even where they harbor serious doubts regarding whether policies being pursued are in the interest of the working class. . .This is a logical development from the situation

whereby most trade unions find themselves in some
form of affiliation or working arrangement with
politicians.[22]

That Mr. Lutchman made these remarks as part of a paper
presented to the TUC at their invitation suggest that some
members of the leadership are uncomfortable with their new
partnership with the Government. Nonetheless, it is
doubtful that this discomfort with an increasingly compromised position will lead to a significant break between
the Government and its allies in the TUM. Prospects for
the development of an independent TUM in the near future
are rather dim because ". . . one of the major problems in
Guyana is the extent to which persons are willing to
sacrifice principle on the altar of expediency, in the
quest for survival."[23]

Moreover, the PNC recently ensured that only persons
willing to toe the party line secured leadership positions
within the TUC. In the Fall of 1980, the ruling party
rigged the TUC elections. PNC affiliated unions showed a
remarkable and impossible increase in membership and the
rules determining the number of delegates allocated to a
union were changed so as to favor the small unions recently organized by PNC activists. The GAWU, the largest
membership union controlled by the PPP, won only one seat
on the Executive Council. The consequences of the electoral changes were immediately apparent.

> The resolutions which were defeated and those
> which were passed made it clear that the future
> of the Guyanese worker is in the hands of the PNC
> regime. Safeguards to protect the worker from
> abuse by the employer, whether in the public or
> the private sector, were swept away.[24]

A motion calling for safeguards to ensure the holding of
"free and fair" elections was rejected by the Delegate
Conference. Another motion condemning the use of the
Security Forces disrupt and hinder the freedom of association guaranteed under the Constitution was similarly
rejected. On the other hand, a motion calling upon the
TUC to give "unstinted support to the Government and . .
.to motivate the workers and farmers to produce in larger
volume and more efficiently . . ." carried easily.[25] It
is then obvious that the established leadership of the TUM
will not provide the nucleus of a movement to challenge
the PNC's corrupt and inefficient rule.

The Degeneration of the PPP.

     The PPP takes great pleasure in pointing out that it
is the true vanguard of the Guyanese working class and
thus the hope for Guyana's socialist future. Whereas the
PNC seized the mantle of marxism-leninism in a desperate
and opportunistic maneuver to retain state power the PPP
asserts that it has for more than thrity years remained
true to struggle for a racially united working class
movement, political democracy and socialist revolution.
Since Dr. Jagan's party maintains the loyal support of the
largest segment of the Guyanese population, it is important that we critically evaluate these claims.
     In 1953 the PPP rose to power with the assistance of
a generally progressive TUM. In less than a decade the
labor movement which the PPP had hoped to shape into
its most powerful weapon against colonialism and capitalism was transformed into its deadly enemy by the American
CIA operating through international labor organizations.
The PPP was left with only the loyal support of an unrecognized union popular among sugar workers. Furthermore, the racial warfare which had devastated the country
in the aftermath of the defeat of the 1963 Labor Relations
Bill had left such a residue of fear and hatred that a
genuine mass-based, multi-racial party was no longer
feasible. The final nail was driven into the coffin
containing the PPP's dream of winning control over the
post-colonial state when the British established the
precedent for electoral rigging with the imposition of
proportional representation. Stripped of a broad base
within the TUM, bi-racial support or a fair chance to win
an election, the PPP degenerated into a parasitic party
subtly appealing to racial loyalty and balatantly living
off the struggles of the sugar workers.
     The PPP maintains the limited political credibility
it enjoys by virtue of its control over the labor force in
a vital economic sector. Whereas the party once sought to
use state power to advance the cause of labor, the PPP now
seeks to use sugar workers to gain a share of state power.
The PNC may arrogantly dismiss the votes of sugar workers,
but it cannot disregard their enormous potential to
disrupt the economy. It is this power of economic disruption versus Jagan's charismatic ability to mobilize the
Indian population which has kept the PPP alive in the
post-independence period. Sugar workers are, moreover,
almost perfectly suited to fulfill the PPP's need for a
political lightning rod. Workers' wages for various field
operations are based on a predetermined, piece-rate.
However, there are daily disagreements between labor

and management over rate adjustments for such things as
poorly burnt cane which makes harvesting more difficult,
pests, or dirty punts. Any one of these minor disputes
has the potential--if either side so desires--for escalating into a full-fledged strike.[26] Therefore, the
PPP has always been able to call a strike whenever its
political strategy so dictated. The crippling sugar
strikes of 1970, 1974-5 and 1977 are excellent examples of this. Thus far the PPP's strategy of economic
disruption has failed to overthrow the Government largely
because the sugar workers--despite their remarkable readiness to sacrifice--are forced to return to work by their
desparate poverty before the PNC is forced to resign.

This political use of the GAWU by the PPP would not
be objectionable were it not for the poor quality of the
party's political and trade union leadership. The GAWU is
completely incompetent when it comes to trade union
affairs. After two decades of industrial struggle, the
GAWU has neither effective collective bargaining nor
grievance handling skills. Consequently, you have an
absurd situation wherein the union most often used to
attack the Government is extraordinarily dependent upon
the Ministry of Labor to carry out its most trivial trade
union functions.[27] Of course, the PPP could argue that
this is the bitter legacy of the Government's and companies' unfair denial of recognition but this conveniently
overlooks its culpability. According to Anthony Haynes,
Bookers' Chief Executive Officer, and Harold Davis,
Chairman of GUYSUCO and former Personnel Director for
Bookers, Jagan always skirted responsibility for the
industrial tasks of the union. For instance, neither the
PPP nor the GAWU ever made presentations before the
numerous commissions of inquiry called to investigate
conditions in the sugar industry and to make recommendations with regard to wages, pensions and profit-
sharing between the time of independence and the GAWU's
recognition ten years later.[28] Furthermore, Haynes,
Davis and Norman Semple contend that Jagan never really
wanted recognition for the GAWU and only accepted it when
it was absolutely necessary. This argument is not as
farfetched as it may seem. Without formal recognition,
the GAWU/PPP was free to strike without the encumbrance of
an institutionalized grievance handling procedure and, as
we have just noted, free from the unglamorous routine of
trade union functions. This was having your cake and
eating it too.

Due to the narrowly political character of the PPP's
goals, the problems of ineffective trade union leadership
persist today. The Industrial Relations Director for

GUYSUCO, Mr. D. P. Sankar, cynically complains that he (a former MPCA officer!) consistently makes more far-reaching demands on behalf of the sugar workers than does the now recognized GAWU. Mr. Sankar is ready to excuse GAWU's lack of collective bargaining and grievance handling skills for historical reasons; what he finds inexcusable is the GAWU's failure to hire professionals and/or train their own people in these vital areas. In a similar vein, the GAWU/PPP has frustrated the efforts of Aston Chase, head of the NAACIE, the clerical union in the sugar industry to organize one industry-wide union. The GAWU/PPP has never flatly rejected the proposition but discussions always break down over the issue of leadership positions and finances.[29] It is interesting to note that Mr. Chase was a prominent member of the early PPP and Minister of Labor in the PPP Government. Today neither Mr. Chase nor the NAACIE is affiliated with the PPP. The reason for this, we believe, is Mr. Chase's continued adherence to the party's original policy wherein holding political office is secondary to advancing the workers' struggle for political and economic democracy rather than the PPP's now inverted approach.

The degeneration of the working class idealism which motivated the PPP leadership in the forties and fifties is today more or less complete. Our analysis strongly suggests that behind the PPP's claims of scientific socialism, lurks a shameless and pathetic contender for state power. Several leading members of the PPP, including PPP parliamentarians and Harry Lall, the former leader of the GAWU, have defected to the PNC. Quite significantly, most defections of PPP legislators occurred around the time of the electoral charades held in 1968 and 1973 when it was painfully clear that the PPP would never come to power by constitutional means.[30] Needless to say, Burnham has made expert use of these apostates as evidence of the correctness of the PNC's brand of socialism and commitment to bi-racialism. This burning desire for public office and the money and prestige it brings is not limited to PPP defectors. Party faithfuls have shown the same concern for personal aggrandizement and organizational security. The PPP's call for a National Patriotic Front Government is excellent evidence of this.

In 1977 the PPP offered the ruling party an olive branch and called for the creation of a National Patriotic Front Government. In the proposal for a new state structure the PPP left unmentioned the issues of (1) corruption and public accountability, (2) the overbearing, lawless bureaucracy and most importantly, (3) the right of organized labor and mass organizations to participate directly in the governance of the country. The kernel of the PPP's proposal was for the creation of a system of two man, joint party leadership. The minority party would appoint the President who would exercise veto power and the majority party would appoint the Prime Minister.[31]   In its critique of the proposal the WPA observed that the PPP had not even bothered to work out a social and economic program on which this proposed coalition of parties would rest.   What was clear, however, were the respective positions for Burnham and Jagan. Jagan, as leader of the party with the support of the majority of the population, would become the country's Prime Minister; Burnham, as head of the minority party, would become the veto wielding President of the Republic.   Burnham, not surprisingly, scoffed at the notion of his taking a back seat position.

He likewise laughed when the PPP at the last minute chose to participate in the 1980 national "elections" although none of their demands regarding free and fair elections had been met.   The PPP defended its decision to contest the elections to their supporters by brow beating them for their supposed fear of armed struggle and inability to even mount a successful political strike.[32] The more convincing explanation of the PPP's decision to contest the obviously rigged elections was the need to reward a few party activists with a seat in Parliament.

What forcefully emerges from the above discussion is the incompetence of the PPP's political and trade union leadership--particularly in the post-independence period. It fumbled its efforts first, to gain control over the postcolonial state by sticking to a doctrinal espousal of "scientific socialism" which made its program appear far more radical than it actually was and encouraged the wrath of its fanatically anti-Communist northern neighbor; second, it failed to develop an effective trade union in the sugar industry or to capture control of the TUC; third, but most importantly, it failed to effectively challenge the emergence of an authoritarian system of government.   Instead, the petty bourgeois leadership of the party retreated from its progressive orientation of the forties and fifties into dogmatic marxism-leninism and loyal membership in the international communist movement.

None of these conclusions, however, detract from the awesome and destructive role external forces have played in the subversion of Guyanese social and political development.   On the contrary, our purpose has been to shed light on some of the ways in which imperialism nurtures and exploits internal weaknesses in progressive movements in the Third World.   In our view, it is long overdue that those committed to political and economic democracy in the

Third World follow the example of the imperialists in this regard and closely scrutinize the flaws in the leftist movements.

Future Prospects

For the time being, the PNC regime has the TUM firmly under its control. Moreover, with 80% of the economy state-owned, this control is, in our opinion, just as important to the maintenance of the PNC's rule as is its expert manipulation of racial hostility. Cooptation, intimidation, official and unofficial violence and the unprincipled exploitation of the GAWU by the PPP all combine to undermine the development of a free trade union movement and democracy in Guyana. Yet it would be premature and unduly pessimistic to conclude that dictatorship has completely triumphed. Productivity continues to drop as the repression of labor increases. Consequently, the PNC's blatant disregard for the workers' demands for democratic unions, an improved standard of living and political democracy is becoming harder to maintain. In 1976, the PNC was forced first to replace the coopted leadership the Guyana Mine Workers' Union and second, to get rid of the MPCA in the sugar industry. Both the GMWU and MPCA leadership had been old and loyal PNC supporters. Through trickery, bribery and violence, the PNC has been able to minimize these victories. The GMWU once again has a leadership hand-picked by the PNC and the GAWU has been maneuvered out of the leadership of the TUC. Nonetheless, each time manipulative and terrorist tactics have to be used to control the labor movement, both the renegade leadership and the Government lose legitimacy. Therefore we fully support Harold Lutchman's conclusion that, "No group should come to believe that it could take the support and posture of the movement for granted . . . If such a situation is allowed to develop there is a danger of the movement being swept aside as an irrelevance and members, or former members, taking matters in their own hands."[33]

The PNC is fixed upon a perilous course. Crude power is self-limiting and the conformity obtained is unreliable. The exercise of power without the softening touch of legitimacy alienates the subject,

> He conforms because of ulterior motives. His conformity is likely to be limited to the matters explicitly backed by power. He will be unlikely

to volunteer information, show initiative, or
co-operate, except when he is explicitly forced
to. Moreover, in moments of crisis, when the
power structure of the organization is weakened,
he will tend to prefer whatever other norms he
subscribes to rather than the organization's.[34]

Hence, Guyana's crisis of productivity. Shortages of
raw materials, energy and spare parts are contributing
factors but labor's sloppy and gruding performance is a
principal concern of management. Fear prevents workers
from directly expressing their opposition to the Government and the established leadership of the TUM but they
make their discontent and defiance known in spontaneous
and informal forms of protest. The PNC's persistent
appeals for increased productivity printed in newspapers,
posted on bill boards and carried over the air waves fall
on deaf ears. The regime has compelled campaign contribu-
tions, attendance at political rallies and educational
courses but it has failed to raise production through
coercion.
    Labor is a conscious, creative process and workers
must therefore choose to give their best. In Guyana while
workers and managers ". . . strive as far as possible to
keep happy those in a position to take action to influence
their career prospects or even their livelihood by strenu-
ously avoiding any suggestion of overt hostility to those
in political control. . .",[35] they also continue to
undermine the economy through poor morale, go-slows,
industrial sabotage and wildcat strikes. Thus the PNC is
confronting a Catch 22. The only way to raise productivity (aside from massive capital intensive investments
which the regime cannot afford) is to allow greater
democracy in the political and economic spheres. This the
party cannot do without running the intolerably risk of
"being swept aside" as an atavistic and repressive structure. The alternative, which the PNC has chosen, is to
go deeper and deeper in debt to the international lending
institutions and consequently increase the repression of
the labor force as required by the IMF's lending policy.
This, in turn, will most likely accelerate the decline in
productivity.[36]
    The remaining question is how long the PNC will be
able to retain state power. Here our conclusions must be
more guarded than many others on the left. According to
James Petras, the contradictions of a state capitalist
regime are so acute as to shortly lead to its overthrow.
Faced with increasing conflict between the workers and the
state, the political elite will, Petras observes, attempt

to fragment working class unity by appealing to racial and ethnic loyalties. This will not, Petras believes, save the regime for long. Socialist consciousness will win out over primordial fears and a revolutionary socialist alliance will seize state power. Petras like many Western leftists pinning their revolutionary dreams on radical regimes in the Third World, is far too sanguine. Guyana's experience thus far has only substantiated the first part of his prediction. Whenever workers have overcome the formidable obstacles to strike action and pushed the regime to the brink of collapse, the PNC has been able to successfully play the race card. In brief, the Black man's fear of "coolie domination" has enabled the PNC since 1968 when it first rigged national elections to maintain the support or at least the complicity of the Black population despite its most flagrant abuse of power.

The proponents of the model of cultural pluralism are likewise only partially correct when they focus upon the racial and ethnic determinants of social processes in Third World societies. There is nothing mysterious or inevitable about their activation or more importantly their persistence. In Guyana, race first took on political significance during the fight for independence and control of the postcolonial state. The imperialist powers and the Burnham forces needed to divide the mass-based movement and they reached for the racial cleaver. Today, both the PNC and the PPP work diligently to stoke the fires of primordial fears in order to maintain their popular base. PNC and PPP speakers never tire of reminding their audiences of the "days of terror" when Blacks and Indians attacked each other's communities with genocidal vengeance. During the most recent national elections, the PNC warned that the "days of darkness" would return if the ruling party should somehow lose the elections. The cover of a PNC campaign booklet left at every house in Georgetown featured a collage of newspaper clippings from 1964 covered with dripping blood. A crude, but no doubt effective reminder of the "days of terror."

In regard to Guyana's immediate future, the prospects for either economic or political progress are not good. In fact, the goal of socialism is not even on the horizon. Ironically, the adoption of marxism-leninism by the PNC has reduced the possibility of such an eventuality. Guyanese people are by now almost completely cynical about the meaning of socialism. From both sides they are inundated with revolutionary rhetoric promising economic development and genuine democracy. In the meantime they spend more and more time on lines waiting to buy a tin of

cooking oil or a bar of soap. The Government proudly
proclaims that health care and education are now free in
socialist Guyana but the citizens know that a patient had
best bring his or her own linen and medicine to the
hospital and that a student is not guaranteed receiving
books or a desk in school. The PPP is fast to criticize
these and other failings of the PNC's rule but then asks
its supporters to participate in obviously fraudulent
elections. Moreover, so long as the PPP is unwilling to
move beyond preaching marxist–leninist revolution they end
up sounding like sore losers. The PNC has cut the ideological ground from under the oppositions's feet as the
Government has adopted one after another of the PPP's
proposals and ultimately its ideological position.
Guyanese simply do not believe that things would have
worked out very differently if the PPP had had responsibility for implementing the revolutionary program.
Furthermore, how can a highly literate Guyanese population
take seriously the PPP's commitment to free trade unionism
and democracy so long as the party takes an uncritical
position in regard to the Soviet Union and the Eastern
bloc countries?

Obviously, marxist–leninist ideology has been used in
the worst sense of the word by the PNC and the PPP to
create an illustion of revolutionary authority; and, in
the case of the ruling party, to legitimize the creation
of an authoritarian state in Guyana. There is, however, a
more serious issue at stake and that is whether or not
marxism–leninism can ever lead the way to socialism in the
Third World even allowing for a detour along the "non-
capitalist path." We doubt that it can. By assigning
the responsibility for socialist revolution to an embryonic class and then entrusting this authority to a party
elite (the state) such theories encourage the emergence of
dictatorship which contradicts the essence of socialism.
It is time for Third World revolutionaries to acknowledge
that they do not know the path that will lead to economic
development and to make a firm commitment to democracy
including its liberal bourgeois manifestations. Those who
argue that freedom of the press, the right to assembly and
free speech are expendible in the struggle for socialism
in the Third World are dead wrong and pointing the way to
petty bourgeois dictatorship.

NOTES

1. Budget Speech 1976.
2. The Trinidad Guardian, 17 March 1977.
3. The Chronicle, 12 August 1977.
4. The Chronicle, 29 October 1977.
5. The Chronicle, 2 November 1977.
6. The Chronicle, 9 December 1977.
7. The Catholic Standard, (Georgetown), 2 November
    1980.
8. Carib News, November 17, 1984; Caribbean Contact, May
    1983.
9. Ministry of Information and Culture, "A Partnership
    with Foreign Investment," Press Release, April 1975.
10. Quoted in the "Debates" 17 November 1977.
11. Desmond Hoyte, Minister of Economic Development and
    Cooperatives, statement made at a press conference,
    10 February 1979.
12. Ministry of Information, "The Guyana Investment
    Code," November 1979.
13. Critics of the Government equate this behavior with
    acts of political sabotage. Mr. Sankar, Chief
    Industrial Relations Manager, GUYSUCO, attributes it
    to indolence and a desire to rip-off any system.
    Whatever the motivation, the corporation must spend
    scare resources to detect and frustrate these
    attempts to hold down productivity.
14. "Proposed Constitution for Worker Participation in
    GUYSUCO," company files, Georgetown, Guyana.
15. Parliament House, "Debates," 19 May 1977.
16. The Government has also built provisions into the
    legislation which will allow it to dilute the three
    voices of labor anytime it so chooses.
17. Interview with D. P. Sankar, Chief Industrial Rela-
    tions Manager, GUYSUCO, Georgetown, Guyana, November
    1980.
18. Parliament House, "Debates," 27 August 1979. (Type-
    written).
19. Quoted in PPP's "Bitter Sugar," n.d., n.p. available
    at Freedom House, Georgetown, Guyana.
20. Eric Huntley, interviewed in Race Today, (September/
    October, 1978) p. 132.

21. Harold Lutchman, "Trade Unions and Human Rights," n.p., August 16, 1980.

22. Lutchman, "Perspectives on the Role of the Trade Union Movement in Guyana," Release, 6 and 7 (1979): 32.

23. Lutchman, "Trade Unions and Human Rights," p. 26.

24. The Catholic Standard, 5 October 1980.

25. Ibid.

26. Interview with Mr. D. P. Sankar, Industrial Relations Director for Guyana Sugar Corporation, Georgetown Guyana, November 1980.

27. Interview with Norman Semple, Chief Labor Relations Officer, Ministry of Labor, Georgetown, Guyana, October 1980. According to Mr. Semple, the MPCA resorted to conciliatory services of his Ministry six times in 1976. On the other hand, "We're snowed under with GAWU conciliations and they're not single issues." In 1977, despite the longest sugar strike in history, GAWU still had to call upon the conciliatory services of the Ministry on nineteen occasions.

28. Interviews with Anthony Haynes, Chief Executive Officer, BMcC, and Harold Davis, Chairman, The Guyana Sugar Corporation, Georgetown, Guyana, November 1980.

29. Interview with Aston Chase, President, National Association of Agricultural, Commercial and Industrial Employees (NAACIE), Georgetown, Guyana, December 1980.

30. Parliamentary Record Book containing membership record, salary scale, and party affiliation of legislators. Available at Parliament House, Georgetown, Guyana.

31. "P.P.P. Calls for a National Patriotic Front," Caribbean Dialogue, 4 (June, 1978): 36-39.

32. Speech by Ram Karran at a "bottom house meeting" held in the sugar belt on November 13, 1980 to explain to PPP supporters why the leadership had decided to participate in the elections scheduled for December 1980.

33. Lutchman, "Perspectives on the Role of the Trade Union Movement," p. 32.

34. Amitai Etzioni, Modern Organizations, (New Jersey: Prentice-Hall, Inc., 1964), p. 51.

35. Lutchman, "Perspectives on the Role of the Trade Union Movement," p. 29.
36. Hilbourne Watson, "The Political Economy of U.S.-Caribbean Relations," Black Scholar, 11 (January/February 1980): 39.

Adamson, Alan H.   Sugar Without Slaves: The Political
    Economy of British Guiana 1838-1904. New Haven: Yale
    University Press, 1972.

Armstrong, Aubrey B., ed. Studies in Post Colonial Soci-
    ety. Yaounde, Cameroon: African World Press, 1975.

Aronowitz, Stanley, The Crisis in Historical Materialism:
    Class, Politics and Culture in Marxist Theory. (New
    York: Praeger Publishers, 1982).

Beckford, George. Persistent Poverty: Underdevelopment in
    Plantation Economies of the Third World. New York:
    Oxford University Press, 1972.

Burnham, Forbes. A Destiny to Mould, Selected Speeches by
    the Prime Minister of Guyana. New York: Africana
    Publishing Corporation, 1970.

Chase, Ashton.   A History of Trade Unionism in Guyana
    1901-68. Demerara: New Guiana Co., 1969.

Cross, Malcolm.   Urbanization and Urban Growth in the
    Caribbean: An Essay on Social Change in Dependent
    Societies. New York: Cambridge University Press,
    1979.

David, Wilfred. Economic Development of Guyana 1953-1964.
    Oxford: Clarendon Press, 1969.

Despres, Leo A.   Cultural Pluralism and Nationalist
    Politics in British Guiana. Chicago: Rand-McNally,
    1967.

Djilas, Milovan.   The New Class: An Analysis of the
    Communist System. New York: Holt, Rinehart and
    Winston, Inc., 1957.

Frank, Andre Gunder.   Capitalism and Underdevelopment in
    Latin America, Historical Studies of Chile and
    Brazil. New York: Modern Reader Paperbacks, 1969.
    _____.   Lumpenbourgeoisie and Lumpendevelopment,
    Dependency, Class and Politics in Latin America. New
    York: Monthly Review Press, 1972.

Furnivall, J.S. Colonial Policy and Practice: A Compara-
    tive Study of Burma and Netherlands India. London:
    Cambridge University Press, 1948.

Girvan, Norman. Corporate Imperialism: Conflict and
    Expropriation. New York: Monthly Review Press,
    1976.

Glasgow, Roy A. Guyana: Race and Politics Among Africans
    and East Indians. The Hague: Martinus Nijhoff,
    1970.

Henry, Zin.   Labor Relations and Industrial Conflict
    in Commonwealth Caribbean Countries. Trinidad:
    Columbus Publishers, Ltd., 1972.

Hoetink, Harry.   The Two Variants in Caribbean Race
    Relations: A Contribution to the Sociology of Seg-
    mented Societies. London: Oxford University Press,
    1967.

194

Hope, Kempe R. Development Policy in Guyana: Planning, Finance, and Administration. Boulder: Westview Press, 1979.

Jagan, Cheddi. The West on Trial: The Fight for Guyana's Freedom. Berlin: Seven Seas Publishers, 1972.

Jayawardena, Chandra. Conflict and Solidarity in a Guyanese Plantation. London: Athlone, 1963.

Kuper, Leo and Smith, M.G., eds. Pluralism in Africa. Berkeley: University of California Press, 1969.

Lewis, Gordan K. The Growth of the Modern West Indies. New York: Modern Reader Paperbacks, 1968.

Lowenthal, David. West Indian Societies. New York: Oxford University Press, 1972.

Mandle, Jay. The Plantation Economy: Population and Economic Change in Guyana, 1838-1969. Philadelphia: Temple University Press, 1973.

Martin, John Bartlow. U.S. Policy in the Caribbean. New York: Westview Press, 1978.

Petras, James. Critical Perspectives on Imperialism and Social Class in the Third World. New York: Monthly Review Press, 1978.

Radosh, Ronald, American Labor and U.S. Foreign Policy. (New York: Random House, 1969).

Rodney, Walter, A History of the Guyanese Working People, 1881-1905. (Baltimore and London: The Johns Hopkins University Press, 1981).

Roxborough, Ian. Theories of Underdevelopment. London: The MacMillan Press Ltd., 1979.

Saul, John. The State and Revolution in Eastern Africa. New York: Monthly Review Press, 1979.

Shivji, Issa G. Class Struggles in Tanzania. New York: Monthly Review Press, 1976.

Smith, M. G. The Plural Society in the British West Indies. Berkeley: University of California Press, 1965.

Standing, Guy and Szal, Richard. Poverty & Basic Needs, Evidence from Guyana and the Philippines. Geneva: International Labor Office, 1979.

Thomas, Clive Y. Dependence and Transformation: The Economics of the Transition to Socialism. New York: Monthly Review Press, 1974.

Young, Crawford. The Politics of Cultural Pluralism. Madison: University of Wisconsin Press, 1976.

ARTICLES

Bellelheim, Charles. "Dictatorship of the Proletariat, Social Classes and Proletarian Ideology." Monthly Review 23 (November 1971): 55-76.

Bitar, S. "The Multinational Corporation and the Relation's Between Latin American and the United States."

Cardoso, F. H. "The Consumption of Dependency Theory in the United States." Latin American Research Review 12 (1977): 7-24.

Cross, Malcolm. "On Conflict Race Relations and the Theory of the Plural Society." Race 12 (April 1971): 477-494.

Despres, Leo. "Differential Adaptation and Micro-Cultural Evolution in Guyana." Southwestern Journal of Anthropology 25 (1969): 19-43.

Dos Santos, Theotonio. "The Structure of Dependence." American Economic Review 60 (May 1970): 231-236.

Grant, C. H. "Company Towns in the Caribbean: A Preliminary Analysis of Christianburg - Wisman - MacKensie." Caribbean Studies 11 (April 1971): 46-72.

_____. "Political Sequence to Alcan Nationalization in Guyana--The International Aspects." Social and Economic Studies 22 (June 1973): 249-271.

Greene, J. E. "The Politics of Economic Planning in Guyana." Social & Economic Studies 23 (June 1974): 186-203.

Henfrey, Colin V. F. "Foreign Influence in Guyana: The Struggle for Independence." in Patterns of Foreign Influence in the Caribbean. ed. by Emmanuel De Kadt. London: Oxford University Press, 1972.

Jacob, C. "Guyana: Victim of Electoral Fraud." Political Affairs 48 (May 1969): 23-28.

Jagan, Cheddi. "Guyana at the Crossroads." Black Scholar 5 (July 1974): 43-47.

_____. "Guyana: A Reply to the Critics." Monthly Review 29 (September 1977): 36-46.

James C. L. R. "The Middle Classes." in Consequences of Class and Color: West Indian Perspectives. ed. by Lowenthal, David and Comitas, Lambros. Garden City: Anchor Press. 1973, pp. 78-92.

Kuper, Leo. "Political Change in Plural Societies: Problems in Racial Pluralism." International Social Science Journal 23 (1971) 594-607.

Lernoux, Penny. "Jonestown Nation." The Nation, November 15, 1980.

Lewis, Arthur W. "The Industrialization of the British West Indies," Caribbean Economic Review 2 (May 1950); 1-61.

Linz, Juan. "An Authoritarian Regime: Spain." in Reader in Political Sociology. ed. by Lindenfeld, Frank. New York: Funk and Wagnalls, 1968, pp. 129-148.

Litvak, I. and Maule, C. "Nationalization in the Caribbean Bauxite Industry." International Affairs 51 (January 1975): 43-59.

Lutchman, Harold. "The Co-Operative Republic of Guyana." Caribbean Studies 10 (October 1970): 97-115.

_____. "Perspectives on the Role of the Trade Union Movement in Guyana." Release 6 and 7 (1979).

Mandle, Jay. "Continuity and Change in Guyanese Underdevelopment." Monthly Review 28 (September 1976): 37-50.

_____. "Problems of the Noncapitalist Path of Development in Guyana and Jamaica." Politics and Society 7 (1977): 189-97.

Magdoff, Harry. "Is There a Non-Capitalist Road?" Monthly Review 30 (December 1978): 1-10.

Newman, Peter. "Racial Tension in British Guiana." Race 3 (May 1962): 31-45.

Ohiorhenuan, John, F. E. "Dependence and Non-Capitalist Development in the Caribbean: Historical Necessity and Degrees of Freedom." Science and Society 43 (Winter 1979-1980): 386-408.

Prescod, Colin. "Guyana's Socialism: An Interview with Walter Rodney." Race & Class 18 (Autumn 1976): 109-128.

Paterson, Orlando. "Context and Choice in Ethnic Allegiance: A Theoretical Framework and Caribbean Case Study," in Ethnicity: Theory and Experience ed: N. Glazer and B.P. Moynihan, (Cambridge, Massachussets: Harvard University Press, 1975).

Premdas, Ralph. "The Rise of the First Mass-Based Multi-Racial Party in Guyana." Caribbean Quarterly 20 (September-December 1974): 5-20.

Przeworski, Adam. "Proletariat into a Class: The Process of Class Formation from Karl Kausty's The Class Struggle to Recent Controversies," Politics and Society, 7 No. 4 (1977)

_____. "Guyana: Socialist Reconstruction or Political Opportunism?" Journal of International Studies and World Affairs 20 (May 1978): 133-164.

Reno, Philip. "The Ordeal of British Guiana." Monthly Review 16 (July-August 1964): vii-132.

Rodney, Walter. "Internal and External Constraints on the Development of the Guyanese Working Class." The Georgetown Review 1 (August 1978): 4-22.

Smith, Raymond T. "Race and Political Conflict in Guyana." Race 12 (April 1971): 415-427.

Smith, Tony. "The Underdevelopment of Development Literature: The Case of Dependency Theory." World Politics 31 (January 1979): 247-288.

Sweezy, Paul. "Modern Capitalism." Monthly Review 33 (June 1971): 1-10.

_____. "On the New Global Disorder." Monthly Review 30 (April 1979): 1-9.

_____. "The Transition to Socialism." Monthly Review 23 (May 1971): 1-16.

_____. "Toward a Program of Studies of the Transition to Socialism." Monthly Review 23 (February 1972): 1-13.

Thomas, Clive Y. "Sugar Economics in a Colonial Situation: A Study of the Guyana Sugar Industry." Studies in Exploitation 1 (n.d.).

_____. "Meaningful Participation: The Fraud of It." in The Aftermath of Sovereignty: West Indian Perspectives. ed. by Lowenthal, David and Comitas, Lambros. Garden City: Anchor Press 1973, pp. 350-360.

_____. "Bread and Justice: The Struggle for Socialism in Guyana." Monthly Review 28 (September 1976): 23-35.

_____. "Guyana's Sugar: History and Development." The Georgetown Review 1 (August 1978): 23-56.

Watson, Hilbourne A. "The Political Economy of U.S.-Caribbean Relations." The Black Scholar 11 (January/February 1980): 30-41.

Wolpe, Harold, "Theory of Internal Colonialism: the South African Case," in Beyond the Sociology of Development, ed: Ivan Oxall, Tony Barnett and David Booth, (London: Routledge and Kegan Paul, 1975).

West, Katharine. "Stratification and Ethnicity in 'Plural' New States." Race 13 (April 1972): 487-496.

Zimbalist, Andrew. "Synthesis of Dependency & Class Analysis." Monthly Review 32 (May 1980): 27-31.

FOREIGN NEWSPAPERS

The Chronicle. Georgetown, Guyana.
The Guyana Graphic.
The Catholic Standard. Georgetown, Guyana.
The Argosy. Georgetown, Guyana.
The Trinidad Guardian.
The Caribbean Contact.

GOVERNMENT, PARTY and CORPORATE DOCUMENTS

Georgetown, Guyana, Parliament House, "Debates." (Typewritten) 1970-1980.

Georgetown, Guyana, Minister of Finance, "Budget Speech." 1974-1977.

Georgetown, Guyana, Ministry of Information, "The Guyana Investment Code." November 1979.

_____. "A Partnership with Foreign Investment." Press Release, April 1975.

_____. "Guyana, A Decade of Progress." December 1974.

Hoyte, Desmond. Minister of Works and Communication.
"Trade Unionism and the State in Partnership or in
Conflict." Critchlow Lectures Series, Publication
No. 3 (August 1973).

Burnham, Forbes. "The Year of the Breakthrough." Third
Anniversary Speech of the Co-operative Republic. 23
February 1973.

_____. "Declaration of Sophia: A New Road--A New
Code to Socialism." Speech at a Special Congress to
mark the tenth anniversary of the PNC in Government.
December 1974.

_____. "Onward to Socialism." Fifth Anniversary
Speech of the Co-Operative Republic. 23 February
1975.

_____. "Towards the Socialist Revolution." Speech at
the First Biennial Congress of the PNC. 18 August
1975.

Peoples' Progressive Party. "Nationalization and the New
Investment Strategy." n.d., n.p. Available at
Freedom House, Georgetown, Guyana.

_____. "Bitter Sugar." n.d., n.p. Available at Freedom
House, Georgetown, Guyana.

_____. "The Case of the Sugar Workers." n.d., n.p.
Available at Freeedome House, Georgetown, Guyana.

Working Peoples' Alliance and the Working People of
Linden. "The Peoples' National Congress versus the
Bauxite Workers." January 1977, n.p.

Booker McConnell Limited, Annual Report and Accounts
1939-1980.

_____. File of the Negotiations for the Nationalization
of Bookers' Property in Guyana. Company Headquarters, London, England.

UNPUBLISHED MATERIALS

DePeana, George. "As It Is." Trade Union Congress,
Articles first written for broadcast on Radio Demarara, Georgetown, Guyana.

Dowden, Rupert A. "Glimpses into Co-operatives." Available at the National Library of Guyana.

_____. "The Co-Operative Movement in Guyana." Written
for the first anniversary of the Co-Operative Republic. Available at the National Library of Guyana.

Gonsalves, Ralph. "Towards a Non-Capitalist Development
Strategy." Paper presented at the Queens college
Conference on Underdevelopment & Development in the
Black World, 8-10 May 1980.

Lutchman, Harold. "Trade Unions and Human Rights." Paper
presented at the Annual Conference of the CCWU,
Georgetown, Guyana, 16 August 1980.

Payne, Hugh W. L. "Historical Background to Co-Operative Socialism in Guyana." 19 January 1975, available at the National Archives, Georgetown, Guyana.

Scott, Brian. "The Organizational Network: A Strategy Perspective for Development." Ph.D. dissertation, Harvard University, 1979.

INTERVIEWS

Donald Augustin, Chief Economic Adviser to the PNC Government. Interviewed in Georgetown, Guyana, December 1980.

Michael Caine, Chairman, Booker McConnell Limited, Chief Negotiator during the nationalization proceedings. Interviewed at company headquarters, London, England, August 1980.

Keith Carter, former political activist for the Peoples' Progressive Party during the 1950's. Interviewed in Georgetown, Guyana, November 1980.

Martin Carter, Former Minister of Information and Culture in the PNC Government. Interviewed in Georgetown, Guyana, November 1980.

Cleveland, Charran, General Secretary of the Manpower Citizens' Association. Interviewed in Georgetown, Guyana, October 1980.

Aston, Chase, General Secretary of the National Association of Clerical, Commercial and Industrial Employees. Interviewed in Georgetown, Guyana, December 1980.

Harold Davis, Chairman, The Guyana Sugar Corporation. Interviewed in Georgetown, Guyana, November 1980.

Hubert B. Greathead, Guyana Representative, Booker Merchants International Limited. Interviewed in Georgetown, Guyana, September 1979.

Anthony, Haynes, Chief Executive Officer, Booker McConnell Limited. Interviewed at company headquarters, London, England, August 1980.

John Huddart, Former Personnel Director of Non-Sugar Businesses, Booker McConnell Limited. Interviewed in Gerrards Cross, England, August 1980.

Cheddi Jagan, Leader of the Peoples' Progressive Party. Interviewed at Freedom House, Georgetown, Guyana, November 1980.

Ram Karran, General Secretary of the Peoples' Progressive Party and the Guyana Agricultural Workers' Union. Interviewed at Freedom House, Georgetown, Guyana, November 1980.

Ian McDonald, Board Member, The Guyana Sugar Corporation. Interviewed in Georgetown, Guyana, August 1979.

Kevin O'Keffe, Company Secretary, Booker McConnell Limited. Interviewed at company headquarters, London, England, August 1980.

Edgar Readwin, Former Chairman, Bookers' Sugar Estates. Interviewed in West Sussex, England, August 1980.

D. P. Sankar, Chief Industrial Relations Manager, The Guyana Sugar Corporation. Interviewed in Georgetown, Guyana, November 1980.

Norman Semple, Chief Labor Officer, Ministry of Labor. Interviewed in Georgetown, Guyana, October 1980.

Anthony Tasker, Former Public Relations Director for the Booker Group of Companies in Guyana. Interviewed in London, England, August 1980.

Frank Thomasson, Former Personnel Director for Bookers' Sugar Estates, Booker McConnell Limited. Interviewed in London, England, August 1980.

Index

ment with, 83–85, 87; Zaman Ali Advisory Committee Report on, 136–39
Trade Union Council: governmental intervention in, 180–81; neo-colonial alliance and, 100; Peoples' National Congress and, 117–20; recognition poll and, 156
Trade union-political complex, 44
Trades Disputes Bill, 153
Tribalism, 19–20, 22

Undesirable Publications Act, 57
Underdeveloped countries, class-based movements in, 22
Underdevelopment: exploitation and, 9–11; Marxist theory of, 5
Union of Agricultural and Allied Workers, 137–38
United Democratic party, 77, 82
United Force: Booker McConnell's association with, 80; capitalist orientation of, 97; nationalization and, 108, 122; in 1960 election, 83; Peoples' National Congress alliance with, 2, 78, 89

United Nations Declaration for a New International Economic Order, 107
United States: development aid from, 106, 175; intervention by, 82–85, 87

Waddington Commission, 57
West Indies Federation, 80
Working class: color-class hierarchy of, 32–41; history of, 29–45; indentured, 30–32, 34, 36–37, 39; revolutionary potential of, 16. *See also* Bauxite workers; Sugar workers
Working Peoples' Alliance, 114, 116, 117, 119–20
World Bank, 110
World Federation of Trade Unions, 53, 56
World market, 120–22, 127

Young Socialist Movement, 110–11

Zaman Ali Advisory Committee, 136–39